Quabbin: The Accidental Wilderness

Quabbin: The Accidental Wilderness

BY THOMAS CONUEL

Massachusetts Audubon Society   Lincoln, Massachusetts

The Stephen Greene Press   Brattleboro, Vermont

Quabbin: The Accidental Wilderness is the 1981 volume of the Man and Nature series of the Massachusetts Audubon Society, Lincoln, Massachusetts 01773

This book was produced in the United States of America.

It was designed by David Ford and published by The Stephen Greene Press, Fessenden Road, Brattleboro, Vermont 05301.

Library of Congress Cataloging in Publication Data

Conuel, Thomas.
    Quabbin, the accidental wilderness.

    (Man and nature; 1981)
    Includes index.
    1. Quabbin Reservoir (Mass.) 2. Reservoir ecology—Massachusetts—Quabbin Reservoir. I. Title. II. Series: Man and nature (Lincoln, Mass.); 1981. QH1.M27 1981 [TD224.M4] 304.2s [333.78′2′097442] ISBN 0-8289-0457-X (pbk.)   81-7088
                                                                                 AACR2

For my mother

Contents

Acknowledgments

I would like to thank the following people who, at one time or another, helped in the writing of this book: Dave Ashendon, Tom Baron, Rita Barron, Dom DiMatteo, Audrey Duckert, Bill Easte, Dick Forster, Wayne Hanley, Veronica Hayes, Donald Howe, Jr., Robie Hubley, Marcis Kempe, Wayne Petersen, Elenor Schmidt, and Bruce Spencer.

In particular, I wish to thank Warren Doubleday and Jack Swedberg, who contributed greatly to the book, and Eileen Dunne and John Mitchell, without whose editorial guidance and encouragement it could not have been written.

Quabbin: The Accidental Wilderness

1 / Meeting of the Waters

From the lookout tower atop Great Quabbin Mountain, the view to the north is of water, wooded islands jutting from the water, and a perimeter of heavily wooded hills. The trees are mostly maples, oaks, and birches, but dense stands of red pine darken the landscape in spots. The view is unblemished by houses, highways, or utility lines. With binoculars the roads become visible. Most are dirt roads, but some old, crumbled blacktop roads also twist through the woods near Quabbin Reservoir.

A turn towards the west brings into view a church steeple in Belchertown, Massachusetts, ten miles away. Farther away, much farther away, and out of sight but never out of mind, are Boston and its populous metropolitan suburbs, which spawned Quabbin fifty years ago and, some think, have been trying to destroy it ever since.

Say the word "Massachusetts" and the picture that comes to mind is of a heavily industrialized state with its center, its heart, pulsing in the eastern corner. Metropolitan Boston, with its famous universities and hospitals, its perpetual traffic jams, and its more than 2.7 million people, is indeed the hub of Massachusetts and as such dominates the state's politics, its people, and its environment.

But drive west of Boston and the state changes. Drive past Route 128, and twenty miles farther west, past Route 495, and the cars and industry of urban Massachusetts begin to thin out. The roads hold less traffic; the population is less concentrated. It is here in the middle of central Massachusetts, itself almost a state apart from eastern Massachusetts and Boston, that Quabbin Reservoir was built. It is here in the middle of one of the nation's most industrialized states that one of the three largest bodies of fresh water in New England lies surrounded by a surprising wilderness.

Its statistics, like the reservoir itself, are outsized. Quabbin holds 412 billion gallons of water. The Metropolitan District Commission (MDC), which manages Quabbin, is fond of saying that the reservoir is the largest body of untreated drinking water in the country. The waters of Quabbin are of an almost prehistoric clarity and taste, a reminder of what the rest of the state's waters were once, before the industrial age ushered in the era of chemically treated drinking water.

On a map of Massachusetts the reservoir looks like a slaphazard streak of blue, a child's finger-painting across the face of central Massachusetts. Two long fingers of water run north to south, one thin and pointed and the other thick and jagged.

Plunging between the two, though not quite separating them, is the wooded Prescott Peninsula, one of the wildest spots in Quabbin and a focal point in the battle to keep Quabbin a restricted wilderness area. Three Massachusetts counties, containing the towns of Orange, Belchertown, Ware, Hardwick, Petersham, New Salem, Shutesbury, and Pelham, border the reservoir in various degrees of proximity.

Quabbin is a happy accident for Massachusetts. What started off as a land grab by Boston, a city badly in need of clean drinking water by the early part of this century, produced a reservoir, a wildlife refuge, and a superb fishing ground. In one sense, it was all ordained by the collision of two continents 350 million years ago.

Geologists call it the Acadian Orogeny: two land masses, Africa and North America, are thought to have collided. The impact crushed and compacted the rock formations of North America and folded and buckled them like toothpaste squeezed from a tube. In the process, an ocean that had covered the land retreated. Today the hills that surround Quabbin Reservoir are the base of what was once a massive mountain range worn down after many years of erosion, a mountain range that may have been as high as the Alps. All the land around Quabbin is part of several gneiss domes. Volcanoes spewed out hot rocks, which were squeezed into gneiss, a coarse-grained malleable rock. The gneiss formed high hills, and then came a series of glaciers, the last of which was in the geologically recent past, approximately 11,000 years ago. The glacier scooped out and deepened the valleys as it moved but left the gneiss hills standing. These eventually formed the perimeters of the Swift River Valley.

Rivers ran through the broad valley, and the high hills were full of deer and other game. The Nipmuck Indians called the valley "Qaben" meaning "meeting of the waters." White settlers arrived and spread through the area around 1735. The Indians battled for and lost the land, and the newcomers cleared it and began to farm. Towns were incorporated, wars were fought, and monuments erected.

As the years passed, the Swift River Valley changed. Farmers left the land, lured either by the promise of the West or by the hope of an easier life working in the factories that sprang up along the three branches of the Swift River. Small firms began to use the water to power the machinery that produced their goods. The Swift River Valley prospered in a quiet way. Old pictures and paintings show towns with broad streets lined with stately elms and American chestnuts. Homes, churches, and Grange halls were built. And then, in less than three decades, the reservoir was planned and built, and the life of the communities of the Swift River Valley came to an end. The flooding of the Swift River Valley, a mere speck in geologic history, was made possible by upheavals of the earth millions of years ago and made inevitable by the growth of Boston.

One of the attractions of Boston to its earlier settlers was the city's fine drinking water. In the eighteenth century Boston had one of the country's earliest municipal water systems, which drew water from Jamaica Pond and distributed it through wooden pipes. The Charles River, flowing from the south of the state near the Rhode Island border and eventually emptying into Boston Harbor, was the natural and obvious first major water supply for the city.

But by the middle of the nineteenth century, Boston had expanded and the waters of the Charles were dying. The river was on its way to becoming the epitome of foul water and the subject of numerous bad jokes ("the politicians of Boston tend to get delusions of grandeur because they can walk on the waters of the Charles").

By the latter half of the nineteenth century, Boston's need for water had drawn attention towards Lake Cochituate, seventeen miles westward, and towards the Sudbury River, twenty-five miles in the same direction. A regional solution was needed, and in 1895 the Metropolitan Water District was created. The district's first water project was a major one, and a precursor to the construction of Quabbin. The water district's planners selected a low area near the towns of Boylston and West Boylston, about forty miles west of Boston, and began to expand a small local reservoir into what was to become the Wachusett Reservoir. Hundreds of residents of Boylston and West Boylston were evicted from their homes and their property taken by the state. The south branch of the Nashua River was dammed with a masonry dam 205 feet above a ledge foundation, and the 65-billion-gallon Wachusett Reservoir began to fill with water in the first decade of the twentieth century.

But Wachusett Reservoir was never intended to quench metropolitan Boston's growing thirst completely. The Metropolitan Water District brochure relates that "in 1928, farsighted men again dealt with the water shortage and the 20-year process of constructing Quabbin was begun." What happened was that metropolitan water planners once again looked westward to solve their water needs and saw on their maps the Swift River Valley.

It was described in later years by local historians as an agricultural area with modest lumbering, some small manufacturing and grain mills, and a tourist industry catering mostly to New Yorkers who came to fish the ponds. There were four towns in the Swift River Valley, though here local historians sometimes disagree, some preferring to count villages as towns and thus producing counts of five, six, and seven. But the four towns that were officially discontinued, blotted out of existence by state mandate, were Dana, Enfield, Prescott, and Greenwich (pronounced "Greenwitch" by natives of the valley). North Dana, Millington, Coolyville, Packardville, Bobbinville, Nichewaug, and Smith's Village, unofficial villages and towns, are also mentioned on occasion. By all accounts, the Swift River Valley towns were like many other rural American towns in the early decades of the twentieth century: small, independent, and sometimes scorned a little by the busy folk who lived in the cities.

On a U.S. Geological Survey map, the attraction of the Swift River Valley as a reservoir site becomes obvious. It was a low-lying area surrounded by rugged hills, most 500 to 600 feet tall, and needed only a little human handiwork to create a gigantic water supply. Three branches of the Swift River, a tributary of the Connecticut, run into this basin. By impounding the Swift River and the smaller Beaver Brook at the eastern edge of Greenwich and the southern part of Enfield, a huge lake could be created. It was clear to the engineers who came to the Swift River Valley in search of a potential reservoir site that they had found what they were looking for.

2 / The Lost Valley

On a September evening in Barre, Massachusetts, ten miles from the eastern edge of Quabbin Reservoir, the parking lot of the Wildwood Nature Center was crowded with cars. The nature center's main hall, a long rectangular room filled with straight-backed chairs, was darkened, and the shades were drawn. A slide show was in progress and there was standing room only. Warren "Bun" Doubleday stood in front of sixty people and began another lecture on what is commonly referred to as "the lost valley," the four towns that once comprised the Swift River Valley and now lie buried beneath the waters of Quabbin Reservoir.

With a pointer Doubleday tapped an engineering drawing projected on the screen and explained how Winsor Dam and Goodnough Dike were constructed. He described intake tunnels and earthen dams and, in clear, precise language, the enormous engineering effort that went into the construction of Quabbin Reservoir. Next came slides of the Swift River Valley towns. There were slides of Dana, Enfield, Prescott, and Greenwich and their houses, churches, and stores. "This is Nellie Hart's house," Bun narrated, "and here is the old Methodist Church, and this is my grandfather's house." There were slides of the railroad that was known locally as the "Rabbit Run" because it stopped so often that the forty-mile trip from Athol to Springfield often took three and a half hours. And there were pictures of the ponds and lakes of the Swift River Valley.

When the slide show was over, Doubleday rolled a short black and white film. As it ran, he provided a narrative. He, his wife Sigrid, and a pilot climbed into a small airplane and flew over the construction site that became Quabbin Reservoir. After the slides of the Swift River Valley, the movie was a mild jolt. The bulldozed and barren landscape looked as if it had been bombed. Trees after cutting were stacked like matchsticks. Cellar holes gaped where buildings once stood. There was not a house in sight. The land was uniformly devastated.

Bun Doubleday (nobody calls him Warren, and the origin of the nickname is obscure) narrated the home movie in a measured voice that at first reminds one of Maine. But he is not from Maine. He was born and raised in Doubleday Village in North Dana, the village named after his great-grand-father, Nehemiah Doubleday.

A vigorous man in his early seventies who looks considerably younger, Doubleday has close-cropped white hair, bright, alert eyes, and wears dark-rimmed glasses. His stories, like his laugh, are dry and quiet. He is part of a small and ever-dwindling group of original residents of the Swift River Valley towns. When the reservoir was built, 3,500 residents were displaced. Nobody is quite sure how many former residents still remain in the area, but the number is small.

Bun Doubleday's family came to the valley from Connecticut in 1795. Nehemiah Doubleday was co-owner of the Doubleday and Goodman Sawmill on the west branch of Fever Brook in North Dana. The sawmill burned in 1915, but its remains still can be seen on a site about two miles from one of the access roads into Quabbin Reserva-tion. Nehemiah built the house in which he and succeeding generations of the Double-day family lived, and that house was the last to be razed when North Dana was lev-eled in 1938. It was a large, neat colonial with a red barn beside it and two maples in the front yard. During the drought in 1965 the stone steps of the house could be seen protruding upward through the mud flats and low water of North Dana.

Myron Doubleday, Bun's father, ran the town grocery store. He retired at the age of forty-five, when the MDC bought out the store. In those days, Bun Doubleday re-called, a man of forty-five who lost his job could not get another one. The creation of Quabbin Reservoir effectively put his father out of business for life.

The valley that Bun Doubleday's ancestors came to live in at the end of the eighteenth century was part of Narragansett Township Number 4, land granted to veterans of the Narragansett Indian Wars in 1675. Greenwich was the first town incorporated from the township in 1754.

Now the floor of Quabbin Reservoir, Greenwich was once an unusually beautiful town with green hills surrounding broad, flat fields. Located in the northeast corner of Hampshire County, it was divided into two parts, Greenwich Center and Greenwich Plains, each with its own post office. Greenwich Plains was a low, rich, fertile valley. On the heights, at the edge of the plain, were the mountains, Mount Pomeroy, Mount Liz, and Mount Zion. Today the mountains are islands jutting through the waters of the reservoir.

As Greenwich grew, it subdivided. The town of Enfield, settled around 1730, grew from the south parish of Greenwich. By 1816 Enfield was incorporated and named after Robert Field, an early settler. Like Greenwich, Enfield was situated on the for-

mer hunting grounds of the Nipmuck Indians, who had been gradually replaced by the white settlers. The west and east branches of the Swift River cut through Enfield. In later years the river helped power the mills that made Enfield the wealthiest of the valley towns. High hills towered above Enfield. One of these, Great Quabbin Mountain, 500 feet above the floor of the Swift River Valley and 1,000 feet above sea level, was a landmark and is today the lookout summit in the public area of Quabbin Reservation.

Dana was located on the banks of the Swift River in Worcester County. It had been carved from the towns of Greenwich, Petersham, and Hardwick and was incorporated in 1801 and named after Judge Francis Dana of the Massachusetts Supreme Court. Four villages made up the town: Dana Center, North Dana, Storrsville, and Doubleday Village. Scattered through the town were three ponds: Pottapaug, near the center of town, "a large and beautiful sheet of water . . . a resort for fishermen"; Sunk Pond in the southwest corner of town, and Neeseponset-Town Pond in North Dana.

By 1822 the town of Prescott was incorporated. It was a mere sliver of a town bounded by the hills of Mount L, Mount Russ, and Rattlesnake Mountain and by the west branch of the Swift River. Historically, Prescott was dealt a modest dollup of fame in the person of Daniel Shays, who lived in the east parish of Pelham before it was incorporated into Prescott. Shays was born in 1747 in Hopkinton, Massachusetts. He moved to the Swift River Valley and during the American Revolution organized a company and eventually became a captain. He was described as a natural leader, courageous, independent, and ambitious. When the war ended, Shays returned to farming in the Swift River Valley and in 1786, along with his friend Luke Day of West Springfield, led a rebellion of farmers against the repressive debtor laws that allowed creditors to strip farmers of their possessions.

History has looked kindly upon Shays and Luke Day. Both are now regarded as having been hardworking farmers striking out against incompetent government officials who were willing to sit by while war veterans lost their homes and farms to stay-at-home war profiteers. Unfortunately, though the cause was worthy the results were disastrous. The farmers' rebellion at first prevented the sitting of local courts in Northampton and Springfield, thus effectively ensuring that no adverse judgments against farmers could be handed down from those courts. Shays and Luke Day then organized a march on the state armory in Springfield, hoping to capture the armory and its supplies of guns and cannons. That foray ended in a rout when government troops fought back and drove the army of farmers out of the state. His army in ruins, Daniel Shays fled back through Pelham and the Swift River Valley and barely escaped being trapped there by government troops. He slipped away and is said to have lived out the remainder of his life in upstate New York. Shays and his fellow rebels were later pardoned by the government.

When the state built Quabbin Reservoir, it also constructed a road along the western edge of the reservoir and named it the Daniel Shays Highway. One wonders what Shays, a modest man by all accounts, would think if he could know that the memorial to his name is a twenty-mile stretch of two-lane blacktop.

The nineteenth century was a fine time to live in the Swift River Valley. The four towns prospered, and villages sprang up near the towns. The valley was full of farms, grain mills, and small manufacturing firms. In Greenwich there was a gristmill, built in 1745, and an assortment of cottage industries. There were factories in Dana, including the Swift River Box Company, and small industries that churned out straw bonnets, billiard legs, and soapstone fixtures. Prescott was known for its fruit farms, its charcoal kiln, and its soapstone industry. For travelers there were the Conkey Tavern, the Atkinson Tavern, the Quabbin Inn, the North Dana Inn, and the Swift River Hotel.

But the twentieth century was less kind to the valley. By 1910, rumors abounded that the Swift River Valley was destined to become a reservoir for Boston. Property values dropped; businesses folded; and when the rumors became fact, the valley towns succumbed almost meekly to the wishes of the water planners from Boston. Prescott was the first to go. In 1830, 758 people had lived in Prescott. In 1927 only 250 people lived there. After a final town meeting in 1927, the town was turned over to the Metropolitan District Commission, which ran it until its official demise on April 28, 1938.

Enfield, Greenwich, and Dana remained relatively unchanged until the end in 1938.

The *Springfield Morning Union*, in an often-quoted story, reported the demise of Enfield as follows:

"Under circumstances as dramatic as any in fiction or in a movie epic, the town of Enfield passed out of existence at the final stroke of the midnight hour.

A hush fell over the Town Hall, jammed far beyond its ordinary capacity, as the first note of the clock sounded; a nervous tension growing throughout the evening had been felt by both present and former residents and casual onlookers.

The orchestra, which had been playing for the firemen's ball throughout the evening, faintly sounded the strain of Auld Lang Syne . . . muffled sounds of sobbing were heard, hardened men were not ashamed to take out their handkerchiefs."

*   *   *

It is easy to see why the Swift River Valley was considered expendable by state water planners. No major businesses would be ruined, no major highways disrupted, no prominent landmarks buried by the waters of Quabbin. The Swift River Valley was a small, out-of-the-way place, totally lacking the political or financial power that could have saved it. Even to this day MDC officials concede that the greatest problem in building Quabbin Reservoir, and a problem that builders of any future reservoir the size of Quabbin will face, is the tricky emotional problem of moving not just a family or two, but obliterating a whole community with its history and sense of shared lives.

Construction of what was to become Quabbin Reservoir was begun in 1928, completed in 1939, and the reservoir filled for the first time in 1946. The first phase of construction did not involve the Swift River Valley directly. Because Wachusett Reservoir was perilously low, a 12.5 mile aqueduct was built between the Ware River in Barre and Wachusett. A horseshoe-shaped tunnel, twelve feet wide and large enough to drive a truck through, was blasted through underground rock 200 feet deep and then lined with concrete. The first water from the Ware River arrived at Wachusett in 1931, just in time to prevent Wachusett from drying up

completely. By that time, after two dry seasons, Wachusett was 81 percent empty.

Construction of the second tunnel, this one to run westward from Barre to the Swift River Valley, began in 1931. That tunnel is thirteen feet high, eleven feet wide, and ten miles long.

Meanwhile, the state was surveying the Swift River Valley, photographing buildings, acquiring land, and preparing to move cemeteries and build dams. The task of creating a reservoir the size of Quabbin was immense. Metropolitan District Commission engineers, under the direction of Chief Engineer Frank Winsor, tackled the problem of impounding the waters by allowing the valley's topography to determine the shape of the reservoir. At the southern tip of the Swift River Valley, there are two gaps in the hills. The Swift River flowed through one gap on its way out of the valley, and Beaver Brook through the other. The gaps were made by prehistoric rivers that once cut through the valley and the surrounding hills. Glaciers had left over 100 feet of deposition above the bedrock of the old riverbeds. The dams that would impound the waters of Quabbin Reservoir had to be built so that their foundations would reach down through the layers of glacial deposits and rest on the solid ledge of the prehistoric riverbeds below.

First, tunnels were built to divert the waters of the Swift River away from the sites of the dams. Then, concrete caissons were lowered to the riverbed. The first experimental caisson was a block of concrete that was sixteen feet high, nine feet wide, forty-five feet long, and reinforced with steel rods. It had a steel edge and at the bottom contained a hollow chamber six feet high. In that chamber, working with pumps and compressed air, MDC laborers excavated sand, mud, and silt from the riverbed and passed it to the top of the caisson in buckets. As the laborers dug, the caisson settled deeper into the riverbed, eventually coming to rest on bedrock 135 feet below the surface. The caisson was groated with concrete to the rock of the riverbed and formed an impermeable barrier. Other caissons were then lowered onto the first caisson and groated in place. The caissons, when in place, formed a concrete wall nine feet thick.

Winsor Dam and Goodnough Dike were then built above the solid cores of watertight concrete. Both are earthen dams, the materials used in their construction having been excavated from the valley itself. The MDC engineers compacted fifteen feet of fine sand over the tops of the concrete caissons and then built a pool of water over the top of what was to be the dam. They then installed pumps in the pool and dumped measured amounts of soil into a mixing box in the pool. The pumps circulated the water through pipes that could be directed to any area of the construction site. The water, mixed with soil from the mixing box, flowed through the pipes out over the top of the dam, with the coarser soils coming to rest near the edge of the pool, while the finer soils were carried toward the center. The fine soils, which are impervious to water, in this way created a watertight center in the dam.

Winsor Dam and the Goodnough Dike are about three miles apart on the southern, most public edge of the reservoir. Winsor Dam, named in memory of Frank Winsor, who died shortly before the construction of the reservoir was finished, is 2,640 feet long, 170 feet above riverbed, and 295 feet above rock ledge. It contains 4 million cubic yards of earth fill and has a 400-foot spillway.

The Lost Valley: circa 1890–1920

Photos of the Swift River Valley from the Donald W. Howe Collection,
courtesy of the Society for the Preservation of New England Antiquities

The Mill Pond at Enfield. The large brick building is the town hall. Steeple on
the left is the Enfield Congregational Church.

Enfield. The view is up the valley of the west branch of the Swift River.

The Mill Pond at Enfield

Smith's Village, One Mile North of Enfield

Unidentified Farmhouse in the Swift River Valley, Late Nineteenth Century

*The car is a Model T Ford; the place is the Swift River Valley; the
man is unidentified.*

Unidentified Woman, circa 1890

Enfield Center in the Late Nineteenth Century. The view is south towards Belchertown. The building in the center is the Swift River Hotel.

Goodnough Dike is only slightly smaller, 2,140 feet long, 135 feet above the bed of old Beaver Brook, and 264 feet above rock ledge.

The reservoir that stretches behind these dams is 18 miles long, with a water surface area of 38.6 square miles and a shoreline of 118 miles. The distance from Boston, at the eastern edge of the state, to Pittsfield, at the western edge, is 140 miles, only 22 miles more than the shoreline of the reservoir. The maximum depth of the water in front of Winsor Dam is 150 feet, and the average depth of the reservoir eight miles away from the dams is 90 feet. The entire watershed of Quabbin is 186 square miles, supplemented by an additional 98 square miles of watershed in the adjoining Ware River basin.

Of the $65 million appropriated for the construction of Quabbin Reservoir, $41 million was spent on the construction of the dams and tunnels. With the remaining money the MDC built roads and an administration building and two baffle dams to circulate the water from the Ware River through Quabbin. The baffle dams built in Greenwich, now near shaft 12, are simple earth dams that, along with a channel dug by the MDC, circulate the waters from the Ware River northward toward Mount Zion and in the process clean the water of debris. The final figure for the construction of Quabbin was $53 million.

In 1938 a hurricane swept through the valley, flattening stands of trees but doing no damage to the newly constructed dams. By 1939, cleanup crews were finishing with the reservoir, picking up construction and hurricane debris and removing the last structures in the Swift River Valley. The dams were finished and roads were built in and around the administration building in Enfield. The aqueduct from Quabbin to the Ware River was completed, and in the northern part of the reservoir a series of small dams were added to improve circulation in some of the shallow areas of the reservoir.

In July 1939 the waters of the Swift River began backing up and spreading over the low areas of the valley. By 1946 the reservoir was full and Quabbin Reservoir had replaced the Swift River Valley on the map of Massachusetts.

*   *   *

Before Quabbin's waters had begun to cover the Swift River Valley towns, historians and displaced former residents were banding together to preserve a memory of life in the valley. And over the years interest in the valley has grown; it is, after all, one of the few places in this country that have been literally wiped off the map.

The Swift River Valley Historical Society now boasts about 350 members. Well over half of these are younger people with no previous connection to the valley. Since its inception the society has continued to grow and attract new members. It does no promotion and sends out a well-written but modest newsletter only a few times a year. Nevertheless, new members keep joining, attracted, it seems, by the chance to mingle with the original residents of the valley and sort through old memories and old maps. The older members of the society continue to be surprised not only by the numbers of new members but also by the home addresses of some of them. There are dues-paying members from Chicago, Los Angeles, and Seattle.

Bun Doubleday, treasurer of the historical society and one of the sources of valley lore, is much in demand as a lecturer on Quabbin Reservoir and the Swift River Valley towns. He graduated from Worcester Polytechnical Institute with a degree in engineering and found work on the Quabbin construction project during the Depression. His job was to analyze samples of the soil that would go into the earthen dams designed to impound the waters. It is ironic, Bun acknowledged, that one of the last residents of the Swift River Valley should have ended up helping the MDC to bury the valley under water. But there were no angry reactions from neighbors when he and other residents of the valley took jobs with the MDC. It was the middle of the Depression and jobs were hard to find. "Nobody was bitter toward me," Bun recalled. "They all understood. The battle to keep our towns had been fought and lost over a decade before. It was three thousand people arguing with two million people. We didn't stand much chance."

The headquarters of the Swift River Valley Historical Society is located on Elm Street, in North New Salem, in a colonial house built in the early nineteenth century. The house, built in 1816 for William Whitaker, a state representative and later state senator, was purchased by the historical society from the MDC in 1962. Since it acquired the house, the society has been engaged in restoring it and turning it into a museum of Quabbin history and artifacts. The house itself is a monument to an era when houses were built with lavish affection and care. It has seven fireplaces, stenciled stairs, walls, and woodwork, carved mantels, and the original blocked wallpaper. But it is as a repository of the valley's past that the house is most unique. Each room is devoted to a valley town, its history, its families, records, photographs, furniture, and genealogical information.

A separate room, called the Quabbin Room, has been set aside for reservoir information. It is dominated by a huge contour-relief map of the Swift River Valley. The map shows the locations and boundaries of Dana, North Dana, Enfield, Prescott, and Greenwich, with houses and roads included. Photographs of the Swift River Valley and its residents are scattered throughout the building.

The society sponsors occasional outings into Quabbin, three in the spring and three in the fall. Depending on the weather, the excursions usually draw from ten to forty people. Bun Doubleday is almost always there, as well as a small contingent of old-timers from the original towns. Another regular is Audrey Duckert, an English professor at the University of Massachusetts in nearby Amherst and a linguist, wild-plant expert, and collector of Swift River Valley oral histories. She is not an original inhabitant of the Swift River Valley but is official librarian for the historical society. During a walk in the Quabbin woods, she is as likely to discourse on language as Quabbin history, as likely to stop the group to point out British-soldier lichen as she is to spot an old cellar hole. Her collection of tapes is as close to a complete oral history of the valley before the reservoir was built as exists anywhere.

A fellow from Greenwich, named Jack Officer, describes on one tape how he spent the winters of his youth cutting ice from Greenwich Pond for eventual sale in places as far away as New York. Ice harvesting, he recalled, was cold, hard work, not for the less than rugged.

On another tape Lester Hager and Myron Vaughn, all former residents of the valley, recount the tale of Popcorn Snow, a legendary and probably half-mythical valley character, who, when he died after a long life, had himself buried in a zinc coffin with a glass top. He lay in his tomb in Dana for years, resplendent in his best suit and diamond tiepin, and valley residents occasionally stopped by to view him. Finally, someone broke the glass top and stole the diamond tiepin, and old Popcorn Snow's remains, exposed to the air, crumbled into dust. At least that's the story.

Jack Officer's reminisences provide a moving account of how many of the older people felt about being uprooted from the valley in which they had spent their lives. "We figured we'd be there for the rest of our lives. It broke our hearts to go, because we figured we would live and grow and die in that country. To my mind, I couldn't find a better place. Everyone used to hunt and fish and work on the roads. It opens up old wounds to go back and look, and it isn't there."

3 / The Accidental Wilderness

A picture of an eagle hangs on the wall of my office. The bird is either just rising from or just settling onto the ice of Quabbin Reservoir. The picture manages to catch, in sharp black and white, the contrast between the powerful bird, with a fierce glint in its eye and its talons open like grappling hooks, and the space behind it, the winter reservoir, frozen and still with no other sign of life on its white surface.

Jack Swedberg, the man who took the picture, spent the better part of a day—he would be hard pressed to say exactly which day, for he has spent hundreds of them in this way—crouched in a blind on the shore of Quabbin waiting for a chance to photograph the eagle. Eagles are wary, quick to spook, and once scared from a spot by a hint of human presence will not return for days or weeks and sometimes never return. Getting close enough to photograph an eagle is an art laced heavily with patience. And it is an art at which Jack Swedberg excels. He saw his first eagle at Quabbin twenty-five years ago. He was standing on the top of Rattlesnake Mountain, a high peak in the northern section of the reservation, when the eagle came drifting by the rock face below him. Swedberg purchased his first camera shortly thereafter. At that time he was working in the construction industry, but the camera and his love of the outdoors formed a passionate combination that first drew him into wildlife photography as a hobby. Later, as his skills with the camera grew, he was able to turn his hobby into a livelihood. Today he has hundreds of pictures of eagles at Quabbin, along with a striking and varied collection of other wildlife photographs taken there. Swedberg is now chief photographer for the Massachusetts Division of Fisheries and Wildlife and in the course of his official duties spends his days and nights prowling Quabbin with his cameras and shooting photos and films of Quabbin's wildlife and its hills, valleys, and waters.

Swedberg is a big, square-jawed man in his early fifties. He has curly gray hair, dresses in jeans and workboots, and laughs easily. He has a job that frustrated executives often dream about over three-martini lunches.

I met Swedberg for the first time one Indian summer day when he took me on a tour of Quabbin. I had visited Quabbin many times before that day and indeed had lived for several years within ten miles of the main gate. I have always appreciated Quabbin as a natural wonder in Massachusetts, a wild, quiet spot where one can escape the noise and frenzy of this urban state. But Jack Swedberg's Quabbin is different from

mine; to him Quabbin is something more. It is to him, perhaps, what the laboratory is to the chemist, a place of almost infinite possibilities and changing creations.

Swedberg has three 16mm films of Quabbin that he shows for the division of fisheries and wildlife, and his photographs have appeared on television and in national magazines. One of the films is a lovingly photographed, hour-long presentation that Swedberg himself narrates. In it he shows some of the abundant wildlife of Quabbin: owls, flying squirrels, red-tailed hawks, a fawn so young it cannot run, a blue-heron rookery, and an awkward bald eagle slipping on the ice of Quabbin as it makes its way toward a deer carcass. There are other Quabbin animals in the movie: beavers that work year round building and repairing their dams, caught by Swedberg's camera in various seasons and at various chores.

As we walked through Quabbin reservation on that November day, Swedberg kept up a running commentary on the trees, landmarks, and wildlife of Quabbin. "We probably won't see any eagles today," he said to me while we picked our way through a deep swamp adjacent to his eagle blind. "November's too early for eagles at

Quabbin." The blind, constructed by Swedberg and other wildlife staff, appeared to be a large beaver hut from the outside, but inside it was furnished with a bench and shelves and a propane heater. It is from that blind, located far out on Prescott Peninsula on the shores of a beaver pond, that Jack Swedberg takes most of his photographs of eagles.

"The bald eagle needs good winter feeding and Quabbin has that," Swedberg said as we stood outside the blind scanning the blue sky, where a lone crow was winging toward a tall pine. "There's a golden eagle in here too. That's a very shy bird. Some people even think it may have a nest in Quabbin because it has been here five years."

Jack Swedberg makes no secret of where he stands on the question of increased recreation at Quabbin. When he shows his films of Quabbin around the state, he also delivers a low-keyed pitch for the preservation of Quabbin as a wilderness. "It's a great place to do some walking," he told the West Boylston Women's Club after showing the film one day. "There are paved roads, dirt roads, and even grass roads." Briefly, he outlined the history of Quabbin, stressing that it is, after all, primarily a reservoir and that its wilderness opportunities were an afterthought and were made possible only by the limited access and stringent protection a reservoir needs. And then he answered a question from a member of the club who wanted to know why people are not allowed to skimobile or ride horses in the reservation. Though she did not ask it, another question hung in the air: why must all that wonderful land be left unused for the sake of some wild animals? It is a topic on which there is precious little middle ground, and high fortresses of opinion buttress the opposing sides.

Put simply, the issue is this: why not open Quabbin up to increased recreation for the citizens of crowded, sometimes stifling, Massachusetts? Why not let the campers, skimobilers, cross-country skiers, sailors, and horseback riders have access to Massachusetts' only wilderness? After all, don't the resources of the state belong to all its people?

Jack Swedberg frowned when we talked of the demand to open up Quabbin. "There are so many areas like that all around the state," he said. "Why make Quabbin into just another state park? It's a unique area now, a fascinating wilderness area." He paused to consider. "It seems somebody is always trying to create the demand for opening Quabbin. The minute a professional planner becomes involved, he will try to create the demand."

It is often difficult to explain to cross-country skiers, bicyclists, and sailboat owners—all participants in quiet, nonpolluting sports—why Quabbin is best left as a restricted wilderness. "If Quabbin were opened up to cross-country skiers," Swedberg said, "you'd have busloads of people coming from Boston and New York to ski the area, because the attraction of the place is so great. And once you expand recreation at Quabbin for one group, other groups will want the same thing."

Swedberg's favorite photographic subject is the eagle, and it is his passion for this great bird of prey that has often put him on the front line in the battles over opening Quabbin up for increased recreation. "The minute you open Quabbin up to increased recreation, you'll lose the eagles," he said. "I've studied them. I know they won't stay."

Massachusetts is a populous state, with 5.6 million people jammed into 8,257 square miles, but because the population is concentrated in the eastern half, open space is abundant. There are 87 state parks in Massachusetts, totaling 247,932 acres, and most of these parks allow camping, cross-country skiing, sailing, and swimming. Add to this a variety of local and municipal parks and several Audubon wildlife sanctuaries (which have some recreational restrictions), and Massachusetts is far from being the concrete playground pictured by some.

Like any state controversy, the debate on opening Quabbin for increased recreation has spawned committees and studies. But the crux of the matter is rather simple: it is getting harder to be alone in the woods or by the shore. The whine of snowmobiles cutting through the woods on a crisp January morning, while once a rare intrusion, is now commonplace. Camping, once a fairly simple, spur-of-the-moment activity, now seems to require, as much planning, precision, and study of maps as a small-scale military operation.

As Americans take to the outdoors, they are increasingly experiencing a blunt truth: crowds cannot enjoy the wilderness. It is an experience requiring solitude, and the very popularity of the outdoors can doom it. Visitors to the National Park Service doubled in a ten-year span—from 114 million people in 1965 to 228 million in 1975—with resulting damage to the sometimes fragile wilderness areas those visitors had come to see. It is a national dilemma. Wilderness in the United States is gradually being destroyed by the very affection it inspires in a people hungry for authentic experiences.

Quabbin, like wilderness areas elsewhere, has begun to feel the crush of human affection. It attracted more than 310,000 visitors in 1979, including 60,000 fishermen. Officials in the MDC now talk of the pressures they are under: more people every year, increased vandalism, and a reduced budget and staff. Patrolling 85,000 acres is a near-impossibility, made even more so by the nature of the terrain, the isolated hills, valleys, and islands of Quabbin. But so far Quabbin has managed to survive its increased popularity. The southern end of the reservoir, near Winsor Dam and the administration buildings, has absorbed the increased flow of traffic, leaving the rest of Quabbin almost as isolated as it was ten years ago. But as more visitors have crossed the main gate near Winsor Dam and used the hiking trails and recreational facilities available at the end of the reservoir, more voices have begun to ask the question of why Quabbin as a whole should not be opened for public use.

Quabbin's history with regard to its recreational uses is somewhat tangled. It can be seen as a good example of wise use of a wilderness area to provide some limited recreational benefits, or as an example of the state's willingness to let politically organized sporting groups, like fishermen, have their way at the expense of those that are not organized, such as sailors, of whom there is no discernible abundance in central Massachusetts.

When Quabbin filled for the first time in 1946, little, if any, demand existed for more and better recreational facilities for the citizens of the commonwealth. The state was less populated and its residents had other things on their minds: building homes, careers, and families after the trauma of World War II. Few families thought of camping out for a whole weekend or of trudging up and down mountains and trails. Quabbin was unbothered in its early days.

In 1952 things changed somewhat. The state legislature, under pressure from fishermen, expanded fishing at Quabbin and allowed the use of small, ten-horsepower, motorboats. And the cries of wilderness purists, who cringe at the sight of a motorboat, and of sailboat owners, who are still baffled as to why motorboats are allowed on a reservoir but sailboats are banned, opened the first controversy over recreation at Quabbin.

Since 1952, decisions regarding recreation at Quabbin have been made by the MDC. Founded in 1919, the MDC is an offshoot of the Metropolitan Water Commission, the agency that originated Quabbin. A nine-story brick building on Beacon Hill, just around the corner from the State House, houses MDC headquarters. Its location gives an immediate hint at one of the MDC's prime problems: it is a large state agency that is far from immune to political demands and solutions. The responsibility of the MDC is to oversee parks, roadways, and water supplies for the metropolitan area. Quabbin, sixty-five miles to the west, is far outside the MDC's usual jurisdiction, yet MDC police, rather than local police, patrol Quabbin. When decisions affecting a reservoir sixty-five miles away are made in Boston, an occasional problem is inevitable. The decision to allow motorboats on Quabbin's waters is now regarded by some as the MDC's biggest mistake in an otherwise unblemished record

in the administration of Quabbin. But the state's sportsmen were, and still are, an organized group with considerable political clout. In 1976 the state-appointed Quabbin Master Plan Committee looked at recreation at Quabbin and concluded a bit wistfully: "Motorboat fishing . . . is the most intensive recreational use on the Quabbin and adversely affects the water, aesthetics, and quiet, but it is an established use supported by strong sportsmen groups and difficult to curtail."

Over the years the regulations have remained somewhat steady: no hunting, camping, swimming, pleasure boating, or motor-vehicle use of any kind. With the exception of the 16,000 acres of Prescott Peninsula and the 60 islands totaling 3,500 acres, Quabbin is open to the public. Travel in the reservation is restricted to foot, and that is the single greatest reason that the Quabbin wilderness has survived in the middle of urban Massachusetts. The state, however, has made some concessions. The 25,000-acre Ware River Watershed adjoining Quabbin is open to almost every possible recreational use, including hunting, snowmobiling, trailbiking, and cross-country skiing. In recent years the MDC has proposed, but never carried through, plans to expand the fishing area at Quabbin toward the south and nearer the restricted Prescott Peninsula, the winter home of the eagles.

The MDC is not especially enamored of plans to open more of Quabbin to the public. The water planners in the MDC like to boast that Quabbin is the largest body of untreated drinking water in the country, and that increased recreation and use, even by quiet nonpolluters like cross-country skiers, would eventually mean that the reservoir's pristine waters would require extensive chemical treatment. The initial cost of treating Quabbin's water would be $75 million to $100 million, and that cost would be paid by the users of the water in the MDC water district, water users who in general live too far from Quabbin to use it for recreation.

The Quabbin Master Plan Committee, in its 1976 study, noted that it is not a single feature but rather a combination of features—the great size of the reservoir and its watershed area, the game, and the restricted use and relative inaccessibility of much of Quabbin—that makes Quabbin what it is. The committee then went on to grapple with another question: is Quabbin underutilized? It is a charge heard often from the proponents of expanded recreation for Quabbin. For there remains something in the soul of even the most urbanized resident of an urban state that hates to see land just sitting there. It is, I think, a throwback to an earlier era when the American dream was to conquer and subdue the land and wring from it a richer and better life. No matter that a new era is upon us, an era of scarcity and conservation, there is something in the American soul that hates to let a good piece of property "go to waste."

The committee concluded that Quabbin is not underutilized. True, the skimobilers, cross-country skiers, trail bikers, off-the-road-recreation-vehicle owners, sailors, and campers are not allowed to use Quabbin. But the hikers are, and there may be far more hikers in the state than anybody suspects.

The last state report on recreation at Quabbin, the State Comprehensive Outdoor Recreation Plan (SCORP), completed in 1976, included a recommendation that eased some of the recreational restrictions at

Quabbin Park, a public area on the southern edge of the reservoir near the main gate. That recommendation, seen by many as the only sensible solution to the demand for recreation at Quabbin, allowed increased use of only the most public area of Quabbin. But the SCORP team also found, in a survey of the recreational habits of Massachusetts residents, that hiking and walking are the most popular year-round activities in the state. And Quabbin is a hiker's paradise. There is the biting nostalgia of the two-mile hike into North Dana Center, in the western part of the reservation, of walking down a broken blacktop road that once served as the main highway for Dana and North Dana, and of arriving at what was once a town common, the road that circled the common now overgrown with briar bushes and goldenrods, and the cellar holes of the buildings that once stood on the common gaping and overgrown with weeds. There is a sense there, on that weed-clogged forgotten town common, of the people and places and family histories that comprised the town of Dana. And there is the water. The waters of Quabbin are only 500 yards away from the former common, lapping quietly against a small hill.

Old Enfield Road, in the southwestern section of the reservation, is now only one mile long. It was once the main highway linking Belchertown, Holyoke, Springfield, the Swift River Valley, and the towns beyond the valley: Petersham, Orange, and Athol. Its blacktop surface has been roughened by weather and encroaching trees, and it has been flooded in spots by several beaver ponds nearby. At the end of the mile the road runs straight into the reservoir, and if you stand on the shore there you can see across the water to Great Quabbin Mountain, far to the right but easy to spot because of its high observation tower.

Soapstone Hill, 511 feet high in the northeastern corner of the reservation, was once the site of a working soapstone quarry. Chunks of yellow-white soapstone lie about on the hill. Earlier, Indians used the stone to make bowls. On the two-mile hike into what was once Doubleday Village, part of North Dana, the remnants of a sawmill can be found scattered in a meadow near the fast-flowing stream that powered the sawmill before the reservoir was built.

In the public area of the reservation the MDC maintains a series of shorter hiking trails. These trails usually run for a mile or a bit longer and wind in and around the summit and the Winsor Memorial. Hawks Place Trail, which begins at Enfield Lookout and follows a dirt road down to the shore, is a favorite winter trail. For it is from Enfield Lookout that the eagles of Quabbin can often be seen on the ice, feeding in the distance. The lookout, on good winter weekends, often attracts a large crowd of binocular-toting eagle watchers. It is against the law to bother the eagles by venturing out on the ice, and the MDC arrests those foolish enough to do so.

*   *   *

If Quabbin Reservation is a hiker's paradise, the reservoir is a fisherman's delight. "Quabbin Reservoir is a deep, soft-water body with two-thirds of the volume consisting of cold-water habitat." So begins a small pamphlet entitled *A Summary of Eighteen Years of Salmonid Management at Quabbin Reservoir, Massachusetts*. To the uninitiated the news that Quabbin Reservoir has deep, cold waters may well bring a yawn. Not so, however, to fishermen, particularly those willing to spend good money and countless hours trapped in small boats trolling the

water in the hope of an encounter with one of the Northeast's best game fish, the lake trout. Quabbin Reservoir is one of the prime fishing spots for lake trout in the Northeast, a fact attested to by the influx of anglers every year at the three public fishing and boat-launching ramps on the reservoir. Bill Easte, an aquatic biologist for the Massachusetts Division of Fisheries and Wildlife, and a man who has overseen some of the fisheries programs at the reservoir, calls the lake trout the "bread-and-butter" fish of Quabbin. The amount of lake trout taken from Quabbin equals or exceeds that of any other freshwater body in the Northeast, Easte says.

Quabbin Reservoir contains nearly 25,000 acres of water, including deep areas high in oxygen, and sandy shallows. The deep water with its high oxygen content forms an environment that is ideally suited for the propagation of trophy-size fish. The reservoir is oligotrophic, young and free of the detritus that comes with age to most lakes. When the reservoir was being flooded, prior to 1946, the bottom was full of plant life and muddied with topsoils. Within a few years this changed, and the bottom of the reservoir is now classified as rubble and stone, with almost no mud or plant life. In the entire reservoir, according to Bill Easte, there is only one deep water hole lacking the amount of oxygen necessary to sustain salmonids, and it is near shaft 12 in Hardwick.

This fact is significant, for as water temperatures get colder, water usually contains less dissolved oxygen. Bill Easte drew a diagram to explain this to me. He showed a large mixing bowl divided horizontally into three parts. At the top of the mixing bowl the water is warm, 70 to 75 degrees in summer, and well oxygenated. A second layer is deeper and colder, about 65 degrees, but still rich in oxygen. Further down the water is cold, but dissolved oxygen is not plentiful.

In most large lakes and ponds in the Northeast, the only environment conducive to lake trout, rainbow and brown trout, and landlocked salmon (all salmonids, which are coldwater fishes) is the middle layer of water. This factor limits the populations of these species in the Northeast, but Quabbin, because its bottom layer is free from oxygen-consuming detritus, does not have this problem. The salmonids, particularly the prized lake trout, feed and breed in its deep,

oxygen-rich waters, waters that are as cold as 47 degrees.

Quabbin also has a generous supply of warmwater species. Bass, both small-mouthed and large-mouthed, white and yellow perch, chain pickerel, and rock bass are all abundant in different parts of Quabbin.

Warmwater fishes are influenced by the rise and fall of the reservoir's level. The drought of the mid-1960s wiped out numerous shoal areas, the favored spawning ground of the large-mouthed bass. The small-mouthed bass were not as affected and became the dominant bass in the reservoir.

The lake trout is a self-sustaining fish in the reservoir, thanks mainly to the rise of a healthy population of smelt, the prime forage fish for the larger trout. The small-mouthed bass is also self-sustaining. The rainbow trout and the landlocked salmon are not self-sustaining and must be stocked. The landlocked salmon in particular has failed to take hold at Quabbin. The problem with the salmon is simply that not enough are produced in hatcheries to stock a body of water the size of Quabbin, and the landlocked salmon has not reproduced in the reservoir. The brown trout has not reproduced in the reservoir either.

The game fish totals for the 1980 fishing season at Quabbin, as might be expected, are impressive. Among the three most sought-after species in the reservoir, 8,328 lake trout were caught, down from 8,427 in 1979; 3,700 rainbow trout were creeled, up from 3,516 in 1979; and 10,319 small-mouthed bass were caught, up from 9,703 in 1979. High numbers of the warmwater fishes also were caught: large-mouthed bass, chain pickerel, yellow and white perch, rock bass, and brown bullhead. While the catches of most other species remained about the same from 1979 to 1980, the bullhead catch declined. That decline has been part of a steady downward slide for the bullhead during the past five years. In 1975, 23,693 bullhead were taken from the waters of Quabbin. In 1979 that figure had declined over three-fold to 7,355. The 1980 catch was even worse, totaling 3,801 bullhead.

There are two theories about the declining bullhead harvest. The first is fairly simple: since the cost of fishing has increased in recent years, fishermen may be losing interest in the less spectacular fish and concentrating on getting their money's worth by angling for trophy-size fish. The second theory is more ominous: bullhead are thought by some aquatic biologists to be more susceptible to acid rain, and the heavy-metal pollution that often comes with acid rain, than other fish species. If this is true, then the fishing waters of Quabbin could be in jeopardy.

Acid rain, as has been fairly well documented by now, is caused by pollutants released into the air by factories that may be thousands of miles away from the water that is eventually acidified. The acid rain causes the pH level in water to drop, making the water acidic. A pH of 4.5 or less is usually considered fatal for fish. Quabbin has a pH of 6.1 in the spring, but the level is thought to have plummeted below that several times in the past. The reservoir is such a large body of water that a diluting effect often saves it. That is to say, even if a particularly acidic rain were to fall on the reservoir, it might be diluted by the larger volume of acid-free water and cause no real harm. However, with continued exposure to acid rain, the waters of Quabbin will gradually become more acidic. Like many lakes in the Adirondacks, Quabbin will gradually

lose its fish populations if the pH drops below 5.0. If the pH drops to 4.5, there
will no longer be sport fishing at Quabbin. At
that point, the dilution effect will make
lake restoration more difficult. As the pH de-
creases, toxic metals normally bonded to
substrate are released in solution to become
absorbed in gill membranes and other tis-
sues. So far, no major studies have been
done at Quabbin to determine if it is being
affected by acid rain.

The premier fishing spot in Massachusetts
required some fisheries management before
it reached its current status. Prior to 1946,
when the reservoir was filled, the roads to
Quabbin were closed and fishing was banned.
If fishing had been allowed, it would have
been bad. The oxygen content of the bottom
water from 1939 to 1942 fell to nearly
zero as topsoil and plants from the former
towns clogged the reservoir, and it was
not until 1943 that the dissolved-oxygen content of the water began to reach acceptable
levels. Beginning in 1946, fishing was allowed on 46 miles of the reservoir's 118 miles
of shoreline, and boat fishing was permitted
in 1952 from the three MDC boat ramps. Today 65 percent of the reservoir is open for
fishing.

Three separate fisheries-management programs have been in effect at Quabbin at
different points in its history. The first ran
from 1952 through 1960 and involved stocking lake trout and walleye. Lake trout is a
coldwater fish and walleye a warmwater one.
The reservoir, it was thought, was large
enough to provide habitat for both fishes.
However, the lake trout prospered and
the walleye did not. The lake trout was
stocked first in 1952, and during the next five
years almost 300,000 specimens of various
sizes were added. The key to the lake trout's
prosperity was the introduction of the rainbow smelt as a forage fish. The smelt was introduced in 1953 and 1954. It was to become
the single greatest factor in the success of
Quabbin lake-trout populations.

Walleye were first caught in the reservoir
in 1960 and have been caught sporadically
since then, but the failure of the species to
establish itself has been attributed to competition from other warmwater fishes and to
the Quabbin water, specifically the limiting
effects of water that is slightly acidic.

While the walleye foundered, the smelt
stocked as forage for the larger game fishes
flourished to the point that its population
had become too large by 1958. Smelt were

The Swift River Valley: 1939

Photos from the Metropolitan District Commission contract books

Moving the Thayer House from Greenwich

Pushing Down the Old Stone Mill, Enfield

Blowing up the Enfield Dam

Enfield, February 1939. The town hall still stands; the town is gone.

Burning Brush on the West Branch of the Swift River, June 1939

The End at Enfield

The Former Town Common at Greenwich

Last Days of Frank Doubleday's House on the Petersham–North Dana Road

spawning in the tributaries of Quabbin and the fry were returning to the reservoirs in such numbers that they were clogging the water-distribution intake screens. The trout of Quabbin grew quickly on a rich diet of smelt, but by 1959 the MDC had had enough and began a smelt-control program that consisted of applying copper sulfate to smelt eggs. The smelt declined, but so did the prized lake trout.

The second fisheries program for Quabbin began in 1957. Lake trout were being caught in the reservoir then, but it was thought that they were not reproducing there. The solution was to try to establish brook, brown, and rainbow trout to replace the lake trout. A new stocking program got underway between 1957 and 1965. Rainbow and brown trout were stocked heavily. Brook trout were stocked in 1957 but failed to establish themselves and were never stocked again in the reservoir. Some still can be found in Quabbin's tributary streams. The rainbow and brown trout in the reservoir showed rapid gains in weight, which were attributed to their plentiful diet of smelt.

As a means of determining whether or not they were reproducing in the reservoir, stocking of lake trout was stopped between 1958 and 1962. Stocking was resumed in 1963 and continued through 1965, by which time it had been determined that lake trout had begun reproducing in the reservoir in 1961 and were on their way to becoming one of the dominant species there. Meanwhile, the smelt-control program continued. This in turn almost ruined the lake-trout project. By 1964, trout caught in the reservoir contained only small amounts of smelt in their stomachs, a big difference from previous years, and the growth rate of lake trout was declining.

All of which brought about the third fish-management program at Quabbin. Brown trout and rainbow trout, it was decided, were never going to be self-sustaining. It was time to try a different sort of fish: the landlocked salmon. In order to give the landlocked salmon, also a coldwater fish, a good start in the reservoir, the brown and rainbow trout, possible competitors, were not stocked during the seven-year landlocked-salmon program. From the start the salmon program faced a serious drawback in the unavailability of enough landlocked salmon in hatcheries. In 1965, 14,420 spring yearlings were put into Quabbin. The recommended rate for stocking the reservoir was 61,000 spring yearlings. In addition, the landlocked salmon stocked in 1965 and again in 1967 had to contend with the declining smelt population. The smelt-control program, combined with a drought that reached its peak in 1967 and reduced smelt breeding areas, and the increasing lake-trout population feeding off the smelt, led to a drastic decline in the once-abundant smelt. Landlocked salmon failed to establish themselves in the reservoir, and by the early 1970s the stocking program was discontinued.

Smelt hold the key to the success of game fishes like the lake trout at Quabbin. Without smelt the trout do not grow large enough or fast enough to sustain a serious sport fishery at Quabbin. This became evident in 1966 and 1967, when catches of coldwater fish declined as the smelt population declined. At that time fishermen began to complain to the MDC and to the Massachusetts Division of Fisheries and Wildlife; one of the best fishing spots in the Northeast was turning barren. The solution was not complicated. The MDC agreed to allow the reintroduction of smelt provided that rotating screens were installed over the intake pipes in the reservoir. The screens were installed and worked. Smelt no longer clogged the water-intake works. In 1968, smelt were restocked in the tributaries along with trays of fertilized smelt. Smelt were stocked again in 1969, and by 1970 they were back in the reservoir in force and stocking was no longer necessary, though they have never reached the level of their abundance in the early days at Quabbin.

Now, in the fourth decade of fishing at Quabbin Reservoir, the situation has stabilized. Every year for the past several years, 15,000 rainbow trout have been stocked in the reservoir. The rainbow does not breed in the reservoir, so it appears that it will always need to be stocked. Only 2,100 brown trout were introduced into Quabbin in 1978. Brown trout have not been stocked since but may be in the future if a suitable strain of the fish can be acquired. Stocking of the landlocked salmon was given another try in May of 1980, when 17,920 were added to the reservoir. By October of that year fifteen-inch salmon, the minimum legal size, were being caught. The landlocked salmon have never bred in the reservoir in the past, so they too, undoubtedly, will always be what fisheries workers call a "put-and-take" fish.

4 / The Nature of Place

Jack Swedberg is not a hasty man, nor is he given to hyperbole, two facts which contribute to his credibility when he describes how he spotted a mountain lion thirteen years ago in the Quabbin woods. Swedberg tells the story this way: he and fellow wildlife-photographer Dick Smith were driving down a Quabbin road one October morning in 1968 when a mountain lion stepped out in front of them. The animal was six feet long and tawny in color. It stood calmly in the middle of the road for a full minute, long enough for Swedberg and his companion to make a positive identification, and then slipped off into the woods.

What makes Swedberg's story unusual is that mountain lions had long been thought extinct in the eastern United States. Since Swedberg's account, others have reported sightings of mountain lions in Massachusetts, sometimes in the Berkshires, sometimes near Quabbin, and it is not fanciful to conclude that this rarest of cats on the North American continent has found at least a part-time hunting range in the Quabbin wilderness.

When wildlife biologists talk or write about wilderness and wildlife in New England, they inevitably mention Quabbin. Two points are always made: (1) Quabbin has a tremendous variation in habitat, and because of that and its relative inaccessibility, (2) the reservation has an abundance and variety of wildlife unequaled in the region except, possibly, in the wildest tracts of Maine. Included in its 118 square miles are open fields, wet meadows, swamps and bogs, and hardwood and coniferous forests. Of the hardwoods, oak is dominant, a pattern that is typical of hardwood forests throughout New England. But the pine forest at Quabbin is unusual; it contains 3,000 acres of a non-New England species, the red pine.

Every ten years the entire Quabbin forest is inventoried. The types of trees and their age and growth are tallied up. This is the kind of information needed to maintain the delicate balance of the Quabbin forest.

The need to balance water and wildlife in the forest are two topics that frequently arise in the conversation of Bruce Spencer, the Metropolitan District Commission's senior forester at Quabbin. The 55,000 acres that circle the reservoir, and in some cases abut the lands of nearby towns, are Bruce Spencer's charge. "I have three concerns here at Quabbin," Spencer told me one October morning in his office near Winsor Dam, "water production, wildlife protection, and growing trees. I try to balance all three needs."

On a series of U.S. Geological survey maps showing the main body of Quabbin Reservoir and Quabbin Reservation, Spencer carefully marked out a grid. For every half-mile of land on the maps, he drew a red circle representing a one-fifth-acre plot of forest. There were 370 such plots, totaling 75 acres. During the dormant growing season of 1980 and 1981, Spencer checked, measured, classified, and numbered every tree in the sample plots that was over 5 inches DBH (diameter at breast height). In doing so, he inventoried one out of every 800 acres of Quabbin forest, recording the type of tree, its size, and classification (seedling, sapling, pole, sawlog, or nonforest tree), the disturbances, disasters, and weed competition each tree faced, and the kind of silviculture the forest needed. After gathering his forest data, he wrote a report establishing guidelines for cutting, thinning, and planting in the Quabbin watershed for the next ten years.

Spencer's first report on the forest census, completed with the help of other Quabbin foresters in 1972, is seventy-seven pages long, with sixteen tables containing information such as the recommended timber harvest by species, and an eighteen-page section of wildlife proposals for Quabbin Reservation.

"I like the inventory," Spencer told me as we traveled through the northern part of the Quabbin forest one November day. "It gives me a chance to see the whole forest. Silviculture is like agriculture," he said. "Open land returns more water to the reservoir than heavy forest, but if you have too much open land you'll get erosion and soil loss. So it's a matter of balance. You need some large trees, some small trees, some hardwoods, some pines, and some fields." One thing Spencer has decided he does not want at Quabbin is the red pine. Red pine has been called a biological desert by some wildlife experts, for with its heavy crown and the resultant carpet of pine needles on the forest floor, little else prospers.

Planting red pine in the open fields and on the construction scars of Quabbin was one of the first acts of forest management when the construction of the reservoir was finished in 1939. Like many first acts, it failed, so Bruce Spencer has decided to clearcut the red pine gradually, allowing lumber contractors to take the pine and sell it as wood chips and pulp byproducts. As he cuts the red pine, Spencer is also returning parts of the forest to the open fields that once graced the area.

Bruce Spencer is luckier than most foresters, especially where wildlife management is concerned. The work of foresters in most large commercial operations is often reduced, however reluctantly, to controlling wildlife for the sake of a quicker and greater timber harvest. When mentioned in connection with commercial forestry, controlling wildlife usually means poisoning beaver and porcupine and shooting deer. Wildlife that interfere with timber harvests must be destroyed. At Quabbin, according to Spencer, a forester is also a resource manager, and that means considering the welfare of wildlife as well as that of trees. The MDC allows Spencer to practice his brand of forestry unimpeded.

Wildlife blends in with forestry, in Spencer's view. "We don't need hunting," he said, "and we don't need insecticides either. Keep the forest healthy and introduce predators and you'll control the deer, the porcupine, and even the insects. For instance, I'd like to get more bobcat into Quabbin and some resident cougar too. That would control the deer population. Deer love ash and oak, and that will be a problem when I try to plant hardwoods; but I'll clear the fields and get the deer to go there and try to bring in natural predators."

No hunting is allowed anywhere on the

mainlands of Quabbin Reservation, but a few acres of MDC land in the towns of New Salem and Shutesbury are open for hunting. "The bow hunters are nailing up the woods in that area," Spencer told me. "They build platforms to hunt from and in the process drive nails into good trees. They don't have any knowledge of the woods: they don't know the difference between a good tree and a bad tree. They could at least use those portable stands sold for bow-and-arrow hunters."

Bruce Spencer is tall and thin and dresses in jeans and heavy boots. Fifteen years ago, after doing undergraduate work in forestry at the University of Massachusetts in Amherst, eighteen miles away, he came to Quabbin. He was a top forestry prospect, second in his class at the university, and Quabbin was a magnet for him.

Spencer is well read in the history of the northern parts of this continent. He has traveled through the Reindeer Lake region of Canada and retraced the footsteps of the first explorers. As we walked through the Quabbin woods that day, we talked of Canada, of how it is selling its resources to supply the United States, and of the forces that are destroying the environment here and may soon destroy Canada. We talked of the north woods and of the Indian cultures that still survive there, a subject on which Spencer possesses an almost encyclopedic knowledge. Spencer's hero is John Muir, the founder of the Sierra Club, a naturalist, conservationist, and world traveler, who discovered a glacier in Alaska, has a national forest named after him, and managed to travel through Russia, Australia, and India. Muir, like Thoreau, believed in living simply. It is a point of view that Bruce Spencer embraces as he moves about his business at Quabbin.

*   *   *

Beavers, the great agents of change in any forest, fairly swarm over Quabbin Reservation. And where beavers prosper, a cycle of growth and decay that is not unique but is certainly accelerated, inevitably follows. Beavers build dams and create ponds, often in woodlands where no ponds were before. The flooded trees in time die, a process that attracts insects, which in turn attract woodpeckers and other birds. The woodpeckers bore holes in the dead trees, and these in time become nesting places for wood ducks and mergansers, which are attracted to nesting spots near the beaver pond. Other animals also use the dead trees for nesting. Owls, squirrels, and raccoons, especially, favor this habitat.

Life in the waters of the beaver pond also changes. Plankton and insects prosper in the water and then become part of the food supply of frogs, salamanders, turtles, snakes, and fish, which in turn attract larger birds, such as the great blue heron. After a decade or two the trees in the pond not already gnawed down by the beavers succumb to wind and disease and fall down. By this time the beavers have moved on to seek new homes. The original pond changes gradually into a marsh or swamp favored by muskrats. Mink come to the swamp to prey on the muskrats. Several more decades pass, and the swamp fills in. Years later, 100 years or more, perhaps, the marsh fills in and becomes an open meadow, where deer, rabbits, and mice live and where predators such as foxes, hawks, and owls come to hunt.

Otters and fishers live on the edge of the reservoir, and snapping turtles swim in the waters. And in the dry cellar holes and old stone walls scattered throughout the woods are chipmunks, gray and red squirrels, snowshoe hares, cottontails, weasels, raccoons, porcupines, and bobcats. The eastern coyote, nicknamed "the new wolf," is also increasing in number and perhaps the mountain lion that Jack Swedberg saw still ranges through the 85,000 acre wilderness.

Birds of prey are common at Quabbin. Eagles, and hawks, including the red-tail, red-shoulder, broad-wing, marsh, rough-leg, goshawk, and osprey are found there. An angry Quabbin goshawk once took offense at Jack Swedberg's plan to photograph its nest and divebombed him, snatching his hat and bloodying his scalp. A picture, taken by a companion of Swedberg's, showed him hat in hand with blood streaming down his face and caused a mild flurry when it appeared in regional newspapers. Quabbin also has a population of turkey vultures, great-horned owls, barred owls, and screech owls. Peregrine falcons, once common in the state but now rare, occasionally pass through Quabbin on migratory routes. The kestrel, another falcon, is more common at Quabbin.

Any body of water the size of Quabbin is attractive to waterfowl, both for nesting and for rest stops during migration. Wood ducks, black ducks, and hooded mergansers are regular breeders. Green-winged teal and common mergansers are migrants and are often seen. Canada geese use Quabbin during their southward migrations but do not stay long. The common loon, once extirpated in Massachusetts, is making a modest come-back at Quabbin. The size and relative inaccessibility of the reservoir, and consequent lack of disturbance, make Quabbin an ideal nesting habitat for this species. A nesting pair, the first seen in Massachusetts since 1887, was observed at Quabbin in 1975.

Quabbin is being used by personnel of the Massachusetts Division of Fisheries and Wildlife in an attempt to reintroduce the wild turkey into the state, where it once flourished. The bird was reintroduced in 1960 on Prescott Peninsula, and in parts of the Berkshires at a later date. The turkeys have prospered in the seclusion of the peninsula and are now spreading through other parts of Quabbin. They are shy birds and can spot movement at seventy yards, rendering themselves difficult to approach or even see.

Quabbin also is home to a long list of smaller bird species, some common and some not so common. Blue-gray gnatcatchers, yellow-throated vireos, and Eastern bluebirds, for example, are all regular breeders in the reservation.

*   *   *

The wind was strong from the south-southwest and beat the waters of Quabbin into a chop of small whitecaps that broke against the northern shore of the reservoir. We were at shaft 12, gate 43, in Hardwick, near one of the public fishing areas, but it was closed and deserted except for our group on a January morning in 1980. Gray clouds were scudding overhead, pushed by the wind, and blocking the sun. The waters of the reservoir were surprisingly open, though, except for some thin ice near two baffle dams north of shaft 12.

"It will be a long time before we'll ever see Quabbin like this again on a January day," Wayne Hanley said, referring to the absence of ice on the reservoir. Wayne Hanley is the former editor of the Massachusetts Audubon Society magazine and an experienced birdwatcher. In 1979 he came to this spot for the first eagle census in New England and

The Eagles of Quabbin

Photos by Jack Swedberg

Bald Eagle

Bald Eagle

Bald Eagle

Bald Eagle

Bald Eagle

Golden Eagle

Golden Eagle

Bald Eagle

it was cold, he said, so cold that the waters of Quabbin were frozen into one large ice mass. But in January 1980 it was not cold. Windy and brisk, yes, but not the bone-chilling cold of a typical January day in New England. The ground was frozen but bare of snow.

Prior to the 1979 census Jack Swedberg had suggested that shaft 12 would be a likely spot for the census takers to position themselves, and that had proved true. Despite the cold during their vigil at shaft 12 that year, Wayne's group spotted eight eagles, many feeding off deer carcasses on the ice.

When we arrived at shaft 12 that morning in January 1980 for the second annual eagle census, Dick Forster, a naturalist at Massachusetts Audubon Society and another member of the group, had taken out a map of Quabbin to show me our position, which he assured me was still the best, with the exception of that of Jack Swedberg, who was in a helicopter flying over the reservation. From where we stood, Mount Pomeroy was straight ahead, across a mile or two of Quabbin water. It was near Mount Pomeroy that most of the eagles had been seen the year before. We were situated on a grassy, enclosed lip of land that juts slightly into the reservoir. Shaft 12 is on one side of the enclosure and is built on the water's edge, breaking up the field of vision from that point. On one side Mount Lizzie and Prescott Peninsula are visible, on the other side, to the north, two baffle dams and Mount Zion.

I asked Dick if the open water would keep the eagles away. "Oh, we'll see them today, but it'll be tough," he said. "On cloudy days like this they blend in with the ridges."

"It was two hours and forty-five minutes before we saw our first eagle last year," said John Bradley, another Audubon official and the fourth member of our group, as we fanned out around shaft 12, scanning the leaden skies with our binoculars. Shaft 12 is a gray stone building that was empty that day except for some scattered tools and two large trapdoors that lead down into the bowels of Quabbin's water-intake shaft.

We saw our first eagle at 9:22 A.M., an immature bald eagle that glided in from the north against the wind. Wayne and I were facing north with the wind at our backs and binoculars lowered when we saw the bird at the same time. I thought for a moment that it was a large crow, but Wayne made no mistake.

"We've got one," he called out, and Dick and John rushed over from the other side of the shaft.

"It's an immature bald," Dick said, tilting his binoculars toward the bird, which settled for a brief instant on a large pine twenty-five yards away but then glided around our lookout post to the right, head out toward the waters near Mount Lizzie Island.

"It's a very young one," Wayne said, "almost no white on it at all."

Identifying bald eagles, especially immature ones, is not easy. The young bald eagle generally is dark brown with a horn-colored bill. As it matures, it begins to show white plumage on the body and tail. By the time the eagle is four years old, a sub-adult, it has dark brown plumage with a white head, white tail, and the distinctive bare lower leg, or tarsus. The golden eagle, when immature, can be mistaken for a bald eagle. It too is primarily dark brown when young, with a patch of white at the base of the tail and some white on the upper surface of the wings. The golden eagle, however, has a fully feathered lower leg that

gives it a bowlegged appearance when it is standing.

The two species of eagles are more easily distinguished by the time they are adults. The bald eagle has a massive bill, long wings with parallel edges, and unfeathered lower tarsi, and it soars with its wings held in a straight horizontal pattern. The golden eagle has a smaller bill than the bald eagle, shorter, broader wings, fully feathered tarsi, and it glides with its wings above the horizontal, pointing slightly skyward.

At Quabbin the odds are that any eagle sighted will be a bald eagle. Two golden eagles have been reported at the reservoir for over five years, but the great majority of eagles wintering at Quabbin are bald eagles.

We waited, moving about in the wind for another half hour after spotting our first eagle, and then walked over to the road that runs across the baffle dam north of shaft 12. As we walked, Dick Forster kept up a running commentary on birds, beavers, and other wildlife he spotted. At Audubon I have heard Dick referred to as "the man with the magic eye," and in the short stroll through the woods the appellation was proved. According to John Bradley, he was walking with Dick through a field once when Forster suddenly stopped, pointed to an object 100 yards away that John could not see, and then began to run towards it. When John reached the spot, Forster had captured a buck moth, a little-known diurnal species. "It was amazing," John recalled. "He could see the markings on a moth a hundred yards away and I couldn't even see the moth."

As we walked that day, Forster often spotted small birds in the trees and underbrush that the rest of us only saw when he pointed to them. He scanned the sky over Quabbin pointing out mergansers and gulls and an occasional crow. Sometimes the birds moved so quickly in the gray sky that I barely had time to glimpse them, much less identify them. Although he himself does not have the magic eye of Dick Forster, Wayne Hanley theorized that many birdwatchers first become interested in the hobby because of their keen eyesight. Jack Swedberg, the man who seems to spot the most eagles in Massachusetts, also has exceptionally keen eyesight.

We walked from the windy, exposed point near shaft 12 up the paved road and then to the north on a dirt road that leads over the first baffle dam. Ice had formed close to the dam, mostly on the northern side. Beavers had gnawed several trees on the water side of the road, and one large maple had been left standing half gnawed through and looking like a slight push would topple it. We moved out onto the second baffle dam. Ahead was Mount Zion. We stopped, scattered yards apart on the dirt road over the dam, and scanned the sky and the surrounding woods. Jack Swedberg's helicopter came into view to the east, hovering low over the barren, snowless hills. We swung our glasses in the direction of the helicopter, hoping to see it put up an eagle. But the helicopter found no eagles while we were watching and soon disappeared. We moved further along the baffle dam toward Mount Zion. Dick was well in the lead; I was several yards back; and the other two were still further back when we spotted our second eagle. This time it was a mature bald and seemed to materialize from the woods of Mount Zion. We saw it for only a few seconds before it winged away towards the southwest. Dick and Wayne identified it instantly. For my part I was glad the eagle census was not dependent on my identifications: the second time I knew it was an eagle by its flight, but how they could tell it was a mature bald was beyond me.

As we began retracing our steps toward shaft 12, the helicopter passed overhead a second time and we pointed in the direction of the eagle, thinking Swedberg might spot the bird too.

Back at shaft 12 we broke for coffee and doughnuts with the wind tossing the coffee about as we tried to pour it from the thermoses. The coffee tasted wonderful in the cold morning air.

The third eagle spotted at shaft 12 was a mature golden eagle, but only Dick Forster saw it. He had moved off 100 yards to the south, heading toward a small peninsula that juts out toward Mount Lizzie Island. Suddenly he yelled and pointed upward. We could not hear what he was saying, but put our glasses to our eyes and scanned the sky. It was as gray and overcast as before, and the clouds were moving even faster across the bleak background. Dick raced up to the shaft and scanned the sky from there, but the eagle was gone. He spotted it, he explained, far up, going with the wind

this time and headed across our watch post. Wayne Hanley shook his head. "If that eagle got up there with that wind and those clouds, he probably was beyond us before we even heard Dick call."

Later, 100 yards up the paved access road, Dick spotted a red-breasted nuthatch, a species that was not common in the winter of 1980. We gathered and watched the pine where Dick had spied the bird, and it soon appeared, flitting in and out of thin cover.

I spotted the next eagle. Wayne and I tired of the nuthatch after a time and walked back to shaft 12. It was shortly before noon and the wind seemed to be dying down a bit, but the air was still chilly. I was looking out toward a narrow gap in the horizon between Mount Zion and the baffle dams when I saw a large bird glide gracefully over the gap. I put my glasses to my eyes and called to Wayne. He swung his glasses in the direction I was pointing in and identified the eagle. "It's an immature bald, but it's got some white on its back. It's not the one we saw earlier," he said.

For perhaps five minutes we watched the bird as it soared and glided around the perimeters of Mount Zion. Dick Forster and

John Bradley returned, and Dick took out his telescope and focused on the eagle, which was in no hurry to be off. The bird was almost adult, Dick and Wayne agreed, probably three or four years old. We continued to watch the bird for another five minutes and then raced over the baffle dams in Dick's car to get closer to Mount Zion. The eagle obligingly was still in place when we arrived. We were now only about a half mile from the bird. And for the next twenty-five minutes the eagle gave us a show. It glided and swooped in and out of the clouds, moving around the mountain and sometimes off to the western ridges, where the contrast was sharp and our view improved. We stopped to have lunch and still the bird soared.

This eagle was probably waiting for something to happen, Dick told us over the meal. Eagles are opportunists; they seldom kill unless the victim is unwary. "They like dumb birds or dead fish, anything that's not too much trouble," Dick said. He told of the eagle he had seen taking a merganser out of the water at Quabbin. The merganser had been on the water and had put its head under to fish, forgetting about, or not seeing, the eagle above. The eagle dived and with a flash of talons carried the merganser off.

The bald eagle was once a common breeding bird in the Northeast. The New England coast, with its craggy shoreline and numerous islands, was an ideal environment for the species. But the industrial boom that transformed New England in the late nineteenth and early twentieth centuries caused the decline of the eagles. The region's major rivers, once home to breeding eagles, were polluted by industrial waste. The human population in the Northeast grew, seaports flourished, and the eagle was shoved farther and farther away from its breeding grounds along the coast. And finally, the great age of synthetics, the world of modern chemistry pioneered by Germany during the first two decades of this century, reached the shores of the United States and changed life in this country irrevocably. We have reaped the benefits and suffered the consequences of living with chemicals. Wildlife, especially eagles, have suffered.

Pesticides, such as DDT, and industrial chemicals, such as PCBs and mercury, have all been found in large quantities in the eggshells of eagles. These chemical contaminants cause eggshell thinning, the major cause of embryo mortality in eagles, as well as other birds. And the damage progresses with time. Eggshells of Maine eagles examined in 1979 were 25 percent thinner than normal, and mercury levels in those eagles were the highest ever recorded. U.S. Fish and Wildlife Service officials, who produced those sobering statistics, also concluded that the eagle population in the northeastern United States numbers less than 100 pairs and has exhibited low production rates for many years. In short, eagles are declining rapidly in the northeastern United States, despite official efforts to save an endangered species.

A U.S. Fish and Wildlife Service report cites impaired reproduction in the eagle population, caused by environmental contaminants as well as a reduced breeding range, as the major loss in the eagle's battle for survival in twentieth-century America. The breeding range of the American bald eagle now is only a fraction of what it was 100 years ago. Small areas in Ohio, Pennsylvania, and New York are still inhabited by bald eagles, but the major remnant of the eagle

population in the Northeast is in Maine. And the eagles that come to Quabbin Reservoir each winter belong to this remnant.

The U.S. Fish and Wildlife Service report concludes with the statement that willful shooting of eagles is still a major problem, and that the critical breeding population of eagles is now at its lowest level ever. And so eagles head for the great American museum, and I wonder how many people in Massachusetts have ever seen a live eagle, and I wonder also if our grandchildren will know of eagles only as giant stuffed birds occupying the place of honor in the museum's bird room.

I returned to Quabbin for the third annual national eagle census on January 1981. That month the entire New England region was locked into an arctic cold wave. The temperature had climbed to 5 degrees at 9 A.M. when I parked my car outside gate 42. At shaft 12 the waters of Quabbin were solid ice, and a fine, four-inch powder of new snow covered the ground, reflecting the sun in long, undulating patterns. The reservation was desolate. There were four of us again: Dick Forster, John Bradley, Wayne Petersen, a friend of Dick, and I. The morning sky was clear, with a strong sun and high white clouds drifting overhead. Long, bright slashes of sunlight broke through the pines on either side of shaft 12. In the distance the hills of Quabbin, Mount Lizzie, Mount Pomeroy, and Mount Zion shone purple and blue, with hardwoods and pines dotting their snow-covered sides. Again we stationed ourselves near shaft 12 and looked out over the frozen reservoir toward Mount Lizzie, to the south. And again we saw the eagles of Quabbin.

There were four carcasses, deer, we guessed, scattered on the ice within our view. Eagles came and fed on them throughout the day as we watched with binoculars and telescopes.

That day we found and followed fresh coyote tracks across the new snow on the ice near the baffle dam, and with the sun sinking and our day ending, we watched three deer, their white tails held high, bounding through the twilight and across the ice near the public boat-launching ramp not far from shaft 12.

When the eagle census was complete, fourteen bald eagles had been spotted at Quab-bin, ten adult and four subadult birds. The count was down from 1980, but that had been a year with a mild winter and open water at Quabbin. The discrepancy was much discussed at a restaurant in nearby Ware, where our group joined several others who had participated in the eagle census and who were now gathered to compare notes, swap stories, and drink whatever was necessary to alleviate the effects of a day spent outdoors in near zero temperatures. In 1979 eight eagles were counted; in 1980, when the winter was warm, twenty-two were counted, but many of the latter were birds that probably wintered farther south normally but that year took advantage of the mild weather at Quabbin. In 1981, during a harsh winter, fourteen eagles were sighted at Quabbin, a cause for optimism, for the figure represented a rise over 1979 data. Bald eagles were also reported along the Connecticut and Merrimack Rivers, and in Montague, Massachusetts. All of which led the eagle watchers to conclude that bald eagles were slowly but steadily increasing in Massachusetts, thanks in large part to the refuge offered them by Quabbin Reservoir.

5 / The Waters of Quabbin

Boston, the tenth largest city in the United States, sits nestled against the Atlantic Ocean in the easternmost part of Massachusetts, at the end of a water pipeline that begins at Quabbin Reservoir.

Bostonians are worrying about many things in the waning days of the twentieth century: crime is up and so are taxes; property values skyrocket while services decline. With all of this on their minds, it is perhaps understandable that most residents of metropolitan Boston do not spend much time thinking about water. The amount of time and energy one can devote to worrying about the sorry condition of the world is, after all, finite.

It seems likely now that Americans who survived the energy crisis of the 1970s will face another crisis in the 1980s: the water crisis. And that is not a happy thought for a nation that has prospered mightily from the twin blessing of cheap energy and pure water. Water is still commonplace, available, and cheap. But energy was once the same way.

Recently, a study conducted by the federal government looked forward twenty years to what the year 2000 will bring and found, as expected, an alarming array of problems and potential problems. Population will be growing in those parts of the world least in need of more people; food will be scarce for many; energy will be in short supply; and many of the world's forests will have been chopped down for firewood. The study also predicted that in some nations, including the prosperous industrial countries of northern Europe, pure drinking water will be so scarce that it will be regarded as a luxury.

Ocean water makes up 98.4 percent of all the water on earth. Fresh water accounts for the remaining 1.6 percent. About 5.8 million miles of fresh water is spread unevenly across the face of the planet. Some of that fresh water is frozen into glaciers at the North and South Poles, locked away from use.

The world's remaining fresh water is in rivers, streams, lakes, and underground reservoirs or aquifers. Despite its relative scarcity in comparison to salt water, which covers 71 percent of the earth, fresh water until now has been readily available in most parts of North America. Some areas, of course, have been less fortunate than others. In the 1930s, drought spread across the southwestern United States and erosion stripped soil, creating the legendary dust bowls. Southern California has suffered

ongoing water shortages for years, and in some parts of that state it is considered rude to stay overnight at a friend's house and ask to take a shower in the morning. Such are the memories of past water shortages. In Arizona's Pima County, where the city of Tucson is located, population has boomed in the past two decades, and the land has sunk twelve feet. The residents of Pima County are using water from an underground aquifer faster than nature can replace it, and as the water table falls away the land surface sinks down with it.

The East Coast also has felt the fear of a water shortage. The mayor of New York was pictured at a fashionable restaurant recently with an upturned water glass in his hand, a gesture, one assumes, designed to discourage restaurant patrons from asking for the customary glass of water with meals. And in Massachusetts, while nobody thus far has tried to discourage restaurant patrons from drinking water, the headlines of local papers have been drumming out an increasingly familiar story: "Water Shortage Shuts U. Mass Amherst," "32 Communities Face Water Shortages," "High Salt Levels Found in Water in 57 Cities and Towns," and "TCE (trichloroethylene) in 9 Cape and Island Towns' Water Supply."

One of the sadder aspects of the water problem is the amount of fresh water that is polluted. That pollution increases with population is hardly a secret: the Great Lakes in this country are a sorry example. The five great lakes contain one-tenth of the total amount of fresh water in the world, but much of their contents is badly polluted.

Massachusetts has its own share of bad water. The Charles River, the Merrimack River, the Housatonic River, and scores of lesser rivers and streams, all have been dirtied. Still, residents of metropolitan Boston do not have to worry much about water pollution, thanks to the existence of Quabbin Reservoir

The Metropolitan District Commission supplies almost half of the fresh water used in Massachusetts, an average of 165 gallons of water per person per day. Quabbin Reservoir, with a water-surface area of 38.6 square miles, is the source of most of that water. Quabbin has a safe daily water yield of 300 million gallons a day, which means that when more than that amount is drawn

from the reservoir, the water level in the reservoir will fall. Quabbin is currently yielding 315 million gallons of water a day, and the reservoir's water level is falling. It stands now at approximately 86 percent capacity.

Normal precipitation in Massachusetts is 44 inches a year. In 1980, a dry year, only 29.39 inches of precipitation fell, making that year the fourth driest in a 163-year record. After that drought, Quabbin did not refill completely until 1976.

The Swift River, the Ware River, and Fever Brook are the source waters of Quabbin Reservoir. Water enters the reservoir from six points: it flows into the northwest area from the west branch of the Swift River at a spot near Shutesbury, New Salem, and Atkinson Hollow; it flows into the northern tip of the reservoir from the middle branch of the Swift River running from Orange through New Salem; it enters the eastern edge of the reservoir from the east and west branches of Fever Brook flowing on either side of Rattlesnake Hill, in what was once North Dana; and it enters near Dana Center, where the east branch of the Swift River empties into Pottapaug

Pond. The last entry point for water into Quabbin is the Quabbin Aqueduct, where water diverted from the Ware River at Barre, ten miles to the east, flows into the reservoir near the baffle dams in Hardwick. The MDC takes an average of 40 million gallons of Ware River water a day for Quabbin but is forbidden by legislation from diverting the Ware River between June 15 and October 15 of each year.

After water enters the reservoir, it goes through three natural purification processes to prepare it for its departure four years later. These processes, coagulation, filtration, and sedimentation, are techniques for treating polluted water in water-purification plants. They occur naturally at Quabbin because of the reservoir's great size and the time it takes for water to circulate through the reservoir.

The waters of Quabbin are not subjected to any of the elaborate chemical cleaning processes that most water undergoes before it is judged potable. However, Quabbin water is treated with small amounts of chlorine, fluoride, and a lead suppressant before it enters the distribution mains near Boston. The addition of chlorine and fluoride is a preventive measure taken for public health reasons; it is not a water treatment. The lead suppressant is added because parts of Boston, Somerville, and Cambridge have old systems with lead pipes that can leach into the water they carry. The lead treatment is a waste, according to some MDC engineers, who would prefer to see the lead pipes removed.

Water flowing into Quabbin from all sources except the west branch of the Swift River meets and mixes near the center of the reservoir, where a narrow, deep opening between Mount Zion and Mount Lizzie funnels it in a southwesterly direction on the start of its four-year trip around the reservoir.

It flows southwesterly to the tip of Prescott Peninsula and then west past the peninsula before turning northward. Midway up the long finger of Quabbin that lies on the western side of the peninsula, the waters of the west branch of the Swift River mix in, creating a swirl. The flow then reverses itself, heading backward towards the southern end of the reservoir from where it came. Here it squeezes eastward between Great Quabbin Mountain and Little Quabbin

Mountain and passes in front of Goodnough Dike. From there the flow is northward, past Mount Lizzie, over the valley floor of what was once Greenwich, and into the basin of the former Quabbin Lake. At that point, the waters are in front of shaft 12 in Hardwick, the beginning of the Quabbin Aqueduct.

The MDC releases 20 million gallons of water per day back into the original bed of the Swift River beyond Winsor Dam, where the water generates power at a small hydroelectric station before continuing its southerly flow. The Chicopee Aqueduct also originates at the southern end of the reservoir and carries water to the towns of Chicopee and Wilbraham and to the South Hadley fire district, all nonmembers of the MDC water district.

Gravity largely is responsible for the movement of the water on its course toward Boston, that and the seventy miles of tunnels the MDC has built in the past fifty years.

The Quabbin Aqueduct, thirteen feet in diameter, carries water first to the Ware River intake in Barre. When the 40 million gallons of water a day are diverted from the Ware River into Quabbin, the flow is temporarily reversed and the water moves westward from Barre into the reservoir and behind the baffle dams that force it northward toward Mount Zion. Here it mixes with water from the Swift River and Fever Brook and circulates through the reservoir. From the Ware River intake the Quabbin Aqueduct connects with Wachusett Reservoir, and from Wachusett the water flows eight miles southeastward to Marlborough, through either the Wachusett-Marlborough tunnel, built in 1965, or the older Wachusett Aqueduct, built in 1898. At Marlborough the water enters the Hultman Aqueduct and the Sudbury and Framingham Reservoirs, a series of smaller holding reservoirs. The Hultman Aqueduct, built in 1940, runs into the three-mile long Southborough Tunnel. After the Southborough Tunnel, the water again enters the Hultman Aqueduct for an eighteen-mile trip to City Tunnel in Newton. Two other MDC aqueducts flank the Hultman and originate from the Sudbury watershed. To the north is the Weston Aqueduct, which was built in 1903 and carries water 13.5 miles into Weston Reservoir and then into MDC distribution lines; and to the south is the Sudbury Aqueduct, which dates from 1878 and carries water into the Chestnut Hill Reservoir in Boston.

The main artery of the MDC's water delivery system is City Tunnel, a deep-rock tunnel twelve feet in diameter and five miles long that connects the Hultman Aqueduct at the Charles River shaft in Newton with Chestnut Hill Reservoir. From this point the MDC's large distribution mains funnel the water to Boston and its surrounding communities. To help in that job, the MDC built two tunnels running from City Tunnel: the City Tunnel addition, which runs northward, carrying water 7.1 miles to Malden and Boston's northern suburbs; and the Dorchester Tunnel, which runs southward and is one of the MDC's rare but embarrassing failures.

The Dorchester Tunnel was built in 1975. It is a ten-foot deep tunnel cut through rock and runs 6.6 miles from Chestnut Hill through Boston and ends at the Boston-Milton line near Dorchester Lower Mills. The tunnel was designed to improve water distribution south of Boston, where growth and increased water demand have been heaviest. But from the day it was finished, the Dorchester Tunnel had a serious problem: it leaked.

When the tunnel was completed in 1975, it immediately cracked. It was closed, relined, and reopened in 1979. However, it continued to leak, though not as badly as it had in 1975, when it flooded several homes, backyards, and cellars adjacent to it. According to MDC officials, it still leaks between 1 and 2 million gallons of water per day, not a large amount when you are distributing 315 million gallons per day, but no cause for celebration either. The leaks are caused by cracks in the rock formations through which the tunnel was drilled. There are two explanations for those cracks. An engineering firm hired by the MDC said the problem was caused by vertical movement of rocks underneath the tunnel. An MDC geologist said it was caused by horizontal jointing of rocks, a condition in which rock formations fracture and then eventually slip and reposition themselves. In either case, the Dorchester Tunnel was built in a bad spot. The tunnel was built through unstable rock formations and will always be susceptible to cracking. Whether the MDC should have known this before it built the tunnel is open to conjecture, but that it tried unsuccessfully to patch the tunnel and re-open it was an obvious mistake.

"There has been a failure, not once but twice in the design of the Dorchester Tunnel," Thomas Baron, an MDC engineer, told me one afternoon in his office at MDC headquarters in downtown Boston. And what can be done about it? Nothing immediate, according to Baron; "we're just going to have to live with it for a while."

The leaky Dorchester Tunnel is not without its comic side, however. When the tunnel first began leaking in 1976, indignant homeowners confronted with cellars full of water began to sue the MDC for damages. But once the lawsuits against the MDC started, claims that seemed somewhat dubious began to appear rather regularly. After the first rash of lawsuits a judge went out to a Dorchester neighborhood to inspect the damage, in this instance a flooded cellar. The cellar was indeed full of water when the judge arrived, and the homeowner claimed a large dollar figure in losses. Some miscalculation had occurred, however, for the Dorchester Tunnel had been drained and closed for six months prior to the claimed flooding. After a lengthy examination of the subject by the judge, it was determined that the phenomenon of the flooded cellar had been accomplished with a garden hose.

In an era of shortages, however, water leaks are a serious subject. The MDC maintains 350 miles of water lines, some of which inevitably leak. Thomas Baron said that the MDC accounts for 97 percent of its water, and the remaining 3 percent is a reasonable loss.

The more serious problems with leaks occur in the water systems of cities and towns, outside MDC jurisdiction. Many of the local water systems use pipes that have been in the ground for 100 years and were never designed for the loads and stress of modern water-delivery systems. A water study commissioned by the MDC in 1975 found that up to 76 million gallons of water per day was unaccounted for. In Boston alone 40 percent of the water pumped into the city daily was unaccounted for. Not all of this water is lost. Some of it simply is not metered. Municipalities do not usually meter or charge for water used by fire departments, schools, or city hospitals.

Some water goes unaccounted for because of faulty meters. In the mid-1970s it was discovered, for example, that Boston Edison, one of the city's biggest users of water was being underbilled by $49,000 a year because of a faulty water meter. However, some of the water that is unaccounted for is leaking into the ground in Boston, Chelsea, Somerville, and other old cities and towns in the MDC water district. Shortly after the MDC study of 1975, it was discovered that in Brighton and Charlestown the city of Boston was losing 345 million gallons of water per year through leaks, at an annual cost of $85,000. A program of leak detection and repair was instituted with almost immediate results: in 1976 the city was using 150 million gallons of water per day, but by 1980 that total was down to approximately 135 million gallons.

As Boston and its suburbs repair water leaks, the MDC is talking of other projects to produce more water. It is a puzzle to many that the MDC should be planning for new water sources before it is sure that leak detection and repair will not produce the additional water that will be needed in the next decade or two. But the MDC does

not see the issue that way. "Leakage should be a high priority," Thomas Baron admitted. "It hasn't been in the past: the criticism is correct. We need to do everything possible to prevent water loss, but at the same time we need to work for a new water source. The demand for water is rising. We're now using 50 million gallons more a day than we used in the 1960s. If a drought hits, we won't ride it out like we did the last time in the mid-1960s."

Consider the problems of the MDC: it sells water to thirty-four member cities and towns, plus ten nonmember towns that buy MDC water under a special agreement. In its charter the MDC is required to sell water to all cities and towns within a ten-mile radius of the State House and to all cities and towns within a fifteen-mile radius of the State House if water is available. A decade or two ago, when fresh water was plentiful and pollution of ground water was unheard of, the MDC had no problem supplying fresh water to whoever wanted it. But Massachusetts grew, and as it grew, local wells and reservoirs were contaminated by chemicals, road salts, and factory effluents. Cities and towns with bad water, or not enough good water for their growing populaces, turned to the MDC for the solution to their water needs. In 1975 Evelyn Murphy, then secretary of environmental affairs for Massachusetts, placed a ban on adding communities to the MDC water district. Her reason for doing so was that the MDC was already exceeding its safe daily water yield. That ban, however, infuriated several South Shore communities, which had been counting on admission to the water district as the solution to their own increasing water problems.

The MDC cannot take on any new water customers until it first discovers and develops a new source of fresh water. This is why water planners at the MDC for fifteen years have been looking toward the waters of the Connecticut River, a class B, fishable and swimmable river, and thinking of diverting some of those waters into the class A, fishable, swimmable, and drinkable waters of nearby Quabbin Reservoir.

The Connecticut River diversion plan stirs passions like few other environmental issues in Massachusetts. It is a plan the MDC has been fussing with since 1965, and it goes to the heart of the water issues facing the state and brings out all the latent, simmering frustrations residents of western Massachusetts experience in their dealings with eastern Massachusetts. At the very core of the matter is a different concept of the state. In eastern Massachusetts the good of the state is defined as what is best for the majority of its citizens, statistically the residents of metropolitan Boston. But in western Massachusetts that view of the state is anathema. For too long, the argument goes, western Massachusetts has been subsidizing the growth, and often the waste and poor planning, of eastern Massachusetts.

Water divides even the state's environmentalists along an east, west line. Robie Hubley has been battling the proposed Connecticut River diversion almost from the day it was conceived. He is the chairman of the Citizens' Advisory Committee on Water Supply, a part-time employee of the Massachusetts Audubon Society, an independent filmmaker, and one of the unofficial leaders of western Massachusetts' opposition to any plans the MDC has for the Connecticut River. "Look at the history of Boston and its water supply," Hubley said one afternoon in Amherst. "Both expand and continue

Quabbin: 1980–1981

The Waters of Quabbin   Leslie A. Campbell

The View from the Enfield Lookout Tower   Jack Swedberg

Rattlesnake Ledges. The view is towards the southwest, across the former valley of the west branch of the Swift River.   Jack Swedberg

Quabbin Reservoir. Winsor Dam is at the lower right.
Metropolitan District Commission

East of Prescott Peninsula. The former Skinner Hill Road, near Doubleday Village, is visible beneath the surface of the water.   Robert Perron

Former Town Common, North Dana   *John H. Mitchell*

Hitching Post, Town Common, North Dana   John H. Mitchell

Unnamed Brook on the East Side of Quabbin Reservoir   *Jack Swedberg*

to expand. The more water they have, the more they will expand. After they expand, they say that because of growth projections they will need more water by a certain date. Well, of course they will."

Rita Barron is the executive director of the Charles River Watershed Association and a member of the State Water Resources Commission. Like Robie Hubley, she has been involved in state water issues for years. But Rita Barron sees the issue from a perspective not shared by Robie Hubley, whose office in Amherst is seventy-five miles west of Boston. "I hope this isn't becoming an eastern Massachusetts versus western Massachusetts issue," she said. "We cannot afford to be parochial in our views on water. I believe in regional self-sufficiency when possible, but when one area has a demonstrated need and another area has a demonstrated supply, then there has to be sharing. After all, we are supposed to be a commonwealth. But the Connecticut River diversion should be a last resort," she added, "an absolutely last resort."

Edward Shanahan, the editor of the *Daily Hampshire Gazette* in Northampton, Massachusetts, a town on the banks of the Con-

necticut River, wrote a column in the mid-1970s that seems to articulate the feelings of residents in that part of the state toward the Connecticut River diversion and toward sharing resources with Boston. Shanahan's column was headlined "Water Grab for Quabbin." "We think it is fair to ask," Shanahan wrote, "why citizens of the eastern part of the state are entitled to our water, a priceless resource. We are unable to take without challenge coal from the fields of Pennsylvania; we can't cart sand from the beaches of the Cape to use on our riverfront beaches or ponds; we can't chop down trees in state forests to heat our homes and provide fencing for our front yards. Why does the MDC not tap the Charles River or the Merrimack, or dig new underground wells in Plymouth?"

To an extent the MDC has heeded Shanahan's advice. For while the Connecticut River diversion plan is the most spectacular of the water proposals to issue forth from the MDC, there are a host of other plans to provide Boston with water. There is in the works a proposal to divert 20 million gallons of water per day from the Sudbury River, treat the water, and then route it into the

MDC's water distribution systems. The MDC calls the plan "flood skimming" and says it would have no ill effects on the Sudbury River. But environmental groups and towns along the Sudbury River are alarmed by the plan. Some feel, despite the protestations of the MDC to the contrary, that it will harm the river's fragile ecosystem. Others see it as a first step by the MDC in what eventually would become a complete takeover of the Sudbury River.

Other water proposals studied by the MDC include reactivating abandoned or reserve water supplies, drilling new wells into the Plymouth aquifer, detecting and repairing leaks, diverting two tributaries of the Connecticut River (the Tullys and the Millers), and, an admitted long shot, desalinating sea water.

Metropolitan District Commission planners, and many environmentalists, believe the proposal to reactivate abandoned water supplies has potential. A study done for the MDC by the U.S. Army Corps of Engineers found that of forty-six abandoned or reserve water supplies, representing a total yield of 130.5 million gallons of water per day, nine of the sites could be reactivated practically. The cost for the nine sites was estimated at over $55 million and the total water yield at 52.5 million gallons per day. Two of the practical sites were in the Connecticut River Valley, but even discounting these two and their 13 million gallons of water per day, the MDC still could obtain 40 million gallons of water per day from reserve or abandoned water sites.

The proposal to tap the Plymouth aquifer also may have potential. A report by the corps of engineers estimated that that aquifer may be capable of supplying up to 200 million gallons of water per day, a figure nearly equal to half the daily requirement of New York City. To tap it, the MDC would need to dig new wells and reactivate old ones. But tapping the Plymouth aquifer might have crippling drawbacks: it might dry up other bodies of water in the area, an issue that is as yet undetermined by the MDC.

Time seems to be on the side of opponents of the diversion of the Connecticut River. In 1965 two events occurred that almost made the diversion a reality. The first was that Massachusetts, along with the rest of the eastern United States, was in the middle of a prolonged and severe drought. Before the rain came in 1967 Quabbin had fallen to 45 percent capacity, and MDC water planners were searching frantically for alternative water supplies. The second was that Northeast Utilities began building its own pumped storage plant on the top of Northfield Mountain, ten miles north of Quabbin Reservoir. That plant now takes water from the Connecticut River, runs it to the top of the mountain, where it is stored in a reservoir 800 feet above the river and when needed is released back into the river through the power company's turbines. Federal regulations required that the utility encourage multiple uses of its reservoir. For this reason, power company officials, mindful of the water problems besetting the MDC in the mid-1960s, went to the MDC with a proposal. If the MDC would pay for the extra pumping, the utility company would make its reservoir slightly deeper so that the MDC could store water there for diversion into Quabbin Reservoir. In 1967 the MDC asked for, and was granted by the state legislature, $25 million to divert the Connecticut River. By 1970 the state had appropriated another $25 million for the diversion, bringing the spending total to $50 million.

The plan was simple. The MDC would build a ten-mile-long tunnel from Northfield to the northern edge of Quabbin Reservoir. It would transport an average of 70 million gallons of water per day from the Connecticut to Quabbin. One part of the plan the MDC did not overemphasize was that it would take that water only during the spring freshet, an average of about seventy days per year. During those seventy days the MDC would be draining 375 million gallons of water per day from the Connecticut River.

By the 1970s the tide had turned against the diversion plan. Environmental groups that had been silent when the diversion first was proposed were suddenly active in hardnosed opposition. Residents of western Massachusetts who had seemed resigned to the plan in 1967 were outraged by it ten years later. All of this had some effect. In 1978 the state legislature resolved that the diversion be considered only as a last resort. The New England River Basins Commission echoed the sentiment that year.

A year later the corps of engineers, citing lack of local support and disagreement between Connecticut and Massachusetts, withdrew from the plan. All of which has left the diversion in a sort of limbo. However, Robie Hubley still sees the diversion as a real and present threat. An environmental impact study of the diversion is being conducted now and a report is expected in 1982.

It is a fact that no one is absolutely sure what impact the diversion would have on either the Connecticut River or on Quabbin Reservoir. But it is a tribute to the purity of Quabbin's waters and the unspoiled land surrounding those waters that so many are concerned.

Jack Swedberg sees the Connecticut River diversion as the first step toward opening Quabbin up to more recreational use. Swedberg's theory is this: once the waters of the Connecticut are mixed with the waters of Quabbin, the end product that leaves the reservoir to be pumped into Boston will require treatment, and not just a shot or two of chlorine. If the water has to be treated, then there will no longer be sufficient reason to say no to motorboat enthusiasts, campers, and even swimmers and recreational groups. Quabbin as a wilderness will be finished.

Swedberg's theory finds no toehold in Tom Baron's office. Baron, who has worked on the diversion plan for the MDC, said that the only factor making the waters of the Connecticut River grade B quality is turbidity. The river carries a lot of sediment but it is not organically dangerous. The sediment would have ample time to settle in the expanse of Quabbin, so water treatment would not be required. It is true that the sediment would create a silt blanket, Baron admitted, but it would be easily incorporated into Quabbin.

But there are concerns other than siltation. There is the not-so-small matter of Quabbin's fishery. The sea lamprey breeds in the waters of the Connecticut River before heading off to the ocean to mature. Bill Easte, of the Massachusetts Division of Fisheries and Wildlife, believes the lamprey eel would find its way into Quabbin despite MDC proposals to use ozone treatments and electric grids to keep it out. Easte is unsure of exactly how much damage the eel would do to the fish of the reservoir. That it would feed on lake trout is a foregone conclusion. That it would destroy the fishery, as Robie Hubley claims, is possible though not proved.

There are dangers to the fish of Quabbin in addition to the lamprey eel. There is the carp, a resident of the Connecticut River, a bottom feeder, a fish that if it were to get into Quabbin would root around the bottom and cause siltation that would bury the eggs of other fish species present in the reservoir. "Have you ever seen a pond full of big goldfish?" Bill Easte asked. "It's muddy and murky and full of debris from the bottom. Carp would have the same effect as goldfish."

A diversion also probably would impede the migration of shad and Atlantic salmon up the Connecticut River to their breeding grounds. The Anadromous Fishes Restoration Project, as the effort to restore the Atlantic salmon is called, dates back to 1966 and has cost millions of state and federal dollars. Only in recent years have the salmon begun appearing in the river in numbers and making their way up the fish ladders of the various dams on the Connecticut. Oppo-

nents of the diversion say it would disrupt the migration of adult fish and kill the younger fish that would be sucked into the diversion tunnel.

Of more serious concern than the fish that will find their way into Quabbin through the diversion tunnel is the maintenance of oxygen in the water. Bill Easte worries that introducing Connecticut River water, which he called more fertile and biologically active than Quabbin water, into the reservoir would eventually affect the sterility of Quabbin's deep, coldwater holes, the home of the lake trout. The silt and nutrients from the river water eventually could lead to lowered oxygen levels in Quabbin's deep holes. If that happened, the fishing would decline.

The most alarming of the diversion's possible effects is that radioactive water might enter Quabbin via the diversion tunnel. The Vermont Yankee nuclear power plant sits ten miles upstream of the intake pumping-station for the Northfield Reservoir. In 1976 and 1977, officials at the plant admitted that two accidents there, caused by a faulty valve, had dumped 83,000 gallons of radioactive water on one occasion and 100,000 gallons on another into the Connecticut River. Officials at the MDC say that monitors would be installed at both ends of the diversion tunnel to prevent radioactive water from entering Quabbin. Opponents counter that even with monitors a small amount of radioactive water would get into Quabbin before the diversion tunnel was closed. And even a small amount of radioactive material in your drinking water is frightening.

Many fears surround the proposal to divert the old, meandering Connecticut River. Various studies have projected that during the approximately eighty minutes per day the diversion would occur, between 42 and 58 percent of the Connecticut would be sucked from the riverbed and into the diversion reservoir on Northfield Mountain. This daily lowering and raising of the water level in the river would cause erosion of the riverbanks. Robie Hubley likened this to a huge tidal flow in the river. The banks of the Connecticut would eventually cave in, he predicted. Hubley also said the diversion would reduce the flow of the Connecticut and interfere with the river's ability to cleanse itself, thereby increasing pollution.

Officials in the state of Connecticut also worry about the proposed diversion, fearing that a decrease in the river's volume will cause the salt wedge of the Atlantic Ocean to creep up the estuary at Old Lyme. Connecticut already has informed the MDC that it will sue to prevent the diversion, a threat that does not worry the MDC unduly. Julia O'Brien, an MDC planner, believes the courts will side with the MDC, as they did during the first stages of the construction of Quabbin, when Connecticut sued to prevent diversion of the Swift River. But Connecticut's threat of a lawsuit to prevent the MDC from commandeering more water has raised an issue that will be heard many times in the coming decades.

It is a fact that growth follows the availability of fresh water, and one that has been amply illustrated by the first settlements in this country. Is it fair to give all the water and therefore all the potential for growth to Boston and eastern Massachusetts and leave western Massachusetts the poorer? Just how much must neighboring states and rural regions sacrifice for the benefit of large metropolitan areas?

Epilogue

On a May morning of birdsong, blackflies, and budding trees, Bun Doubleday and I went to Quabbin Park Cemetery to walk the graveyard where 7,500 bodies disinterred from thirty-four cemeteries in the Swift River Valley are reburied. A morning fog that had hugged the ground earlier had burnt off and a light breeze stirred. Underneath our feet the grass was thick, newly cut, and free of weeds.

Followed by small swarms of the blackflies, we moved along, crossing the hard-packed dirt roads that circle the cemetery. We paused first in front of the graves of Joseph Doubleday, Bun's uncle from North Dana, and Nehemiah H. Doubleday, Bun's great-grandfather buried beside his wife Abby, and son Ozi. Joseph Doubleday's grave is marked with a simple stone. Nehemiah's is topped by a small, ornate marble monument with four slender columns and a miniature dome, and the marble is discolored in spots. Bun told me a story of a letter he read many years ago that had been written to Abby Doubleday by a spiritualist or medium claiming to have communicated with Abby's son Ozi, who died in 1854 at the age of eight weeks. In the letter Ozi described himself to his mother as being six-teen years old, over six feet tall, weighing two hundred pounds, and living happily in the world beyond the grave. What comfort Abby Doubleday took from the letter is unknown.

The MDC set aside Quabbin Park Ceme-tery for the residents of the Swift River Valley towns. For many residents of the val-ley, this was the final indignity; the disinter-ment and removal of their dead to make way for a reservoir. When the graves were moved from the valley prior to the flooding, the MDC offered a choice. It would rebury the bodies in any cemetery the relatives chose or in Quabbin Park Cemetery. Bun Doubleday's mother, father, and grandfather are buried in Athol, Massachusetts. Bodies are buried randomly throughout the grassy, rolling hills of Quabbin Park Cemetery, with no attempt made to reinter them ac-cording to the original graveyards.

Bun stopped near a simple granite marker with the inscription: "Charles N. Downing 1894–1962, His Wife, Gertrude E. Metcalf 1891–1960." "That is my old friend Charlie Downing," he said. "He was a very good friend, a very good friend. We used to hunt together in Enfield, and later we worked together in the soil laboratory when they were building the reservoir." We moved along and other graves brought other memo-ries. There was the grave of Fred Farly, a surveyor who worked with Bun on the Quabbin construction project, and of Frank Hall, the postmaster of Greenwich.

"A lot of people want to know how they managed to find all the bodies, especially since some of the graves were pretty old," Bun remarked. "I knew the fellows who did the work digging. They used to store the boxes, empty coffins, in the soil-testing lab. They always found some indication of where the actual body had been. They'd find bones or buttons or maybe just discol-ored earth. Of course, they undoubtedly missed the Indian graves."

We circled back to the front of Quabbin Park Cemetery, where a small memorial for war veterans is shaded by the leaves of an old oak. Two flagpoles flank two old can-nons, one from Dana Common and one from Enfield Common. Behind the cannons is a stone monument from Dana Common with a bronze plaque inscribed with the names of Dana soldiers who fought in the American Revolution, the War of 1812, the Civil War, and World War I. A tall stone spire erected in memory of George Washington in 1852 by John Atkinson, a soldier who served under Washington, stands a few feet away. The Enfield Civil War monument, which once stood on the town common in Enfield, is here also. A Memorial Day service is held every year near this spot in the cemetery. It is attended mostly by former residents of the Swift River Valley and their relatives. Several years ago there was talk of discontinuing the service, but the Quabbin survivors protested and it goes on.

The blackflies were becoming bothersome, and the sun had disappeared behind gray clouds as Bun Doubleday and I turned back to the car. I recalled a conversation I had had with Rita Barron at the Charles River Watershed Association. I had mentioned that some water planners in the state thought that three or four additional Quabbins should be constructed in the decades ahead to meet the state's water needs. Rita Barron had replied, "although we will need more regional water systems, it is highly unlikely that anybody will get away with building another Quabbin."

Bun Doubleday agreed; "there was opposition to Quabbin in my day but it wasn't organized. It was just individuals trying to stop the MDC. Today there would be much more opposition and it would be more organized. Today activists from other areas and other states would come into the fight. We wouldn't be fighting alone."

Three days after my walk through Quabbin Park Cemetery, I stopped by the Quabbin administration building. The engineer in charge of records told me that the reservoir was 86 percent full. The residents of Boston and the other cities and towns in the metropolitan water district would not run out of water in 1981.

Outside I drove from the administration building up to the lookout tower on Great Quabbin Mountain. From there I could see the hills bordering the reservoir. In front of me stretched the water of Quabbin. It was for this water that the Swift River Valley was flooded. It was because of this water that the wilderness, with its eagles and its extensive woodlands and abandoned cellar holes, exists in the Quabbin region. From where I stood, the roads that once linked the towns of the Swift River Valley were not visible, only the water and the high lands. A stiff breeze rolled whitecaps across the water's surface, and waves beat against the base of the hill with a steady chop. Overhead, above Great Quabbin Mountain, a red-tailed hawk glided by on a thermal updraft.

Index

Thomas Conuel received
a bachelor's degree in English
from the University of Massachusetts in Amherst
and a master's degree in journalism from Boston University.
He has worked
as a newspaper editor and reporter,
a free-lance environmental writer,
and a technical writer.
He has been a devotee of Quabbin
for over a decade.
The Accidental Wilderness
is his first book.